Getting the most out of your DRILL PRESS

Rockwell Handbooks

GETTING THE MOST OUT OF YOUR DRILL PRESS
GETTING THE MOST OUT OF YOUR TABLE SAW
GETTING THE MOST OUT OF YOUR BAND SAW

Getting the most out of your
DRILL PRESS

a Rockwell Publication

A complete handbook describing the drill press in the home workshop with more than three hundred photographic illustrations and line drawings.

 Rockwell International

Power Tool Division
400 North Lexington Avenue
Pittsburgh, Pennsylvania 15208

Foreword

"Getting The Most Out Of Your Drill Press" is published as a service to power tool users. Because different makes, models, and sizes of machines vary in their performance and features, the editors have tried to make the information in this handbook as general as possible.

The drill press is one of the most basic and useful tools to be found in the home workshop. It is also one of the safest. But, as with any power tool, certain safety precautions must be followed. These procedures are included throughout the book and are detailed in Chapter 2.

Drill presses are available in a number of sizes and designs. When selecting a machine, make sure it will meet your requirements. If you intend to use your drill press for long, sustained periods of time, you should consider buying a unit with a greater capacity and heavier duty construction. Also, be certain that the machine you select can be adapted for all of the operations you will perform on it.

This book was designed to accommodate both the beginner, wanting to learn the basic operations of the drill press, and the skilled artisan, needing information about a single, more complex, area of drilling.

Originally published in 1937, this handbook has been reprinted many times, and has evolved into both a compendium and a final word about drill presses. This most recent edition includes all applicable information from previous editions, as well as the latest in woodworking and metalworking techniques. Material for this book was based on information gathered from sources throughout the power tool industry. It is the greatest hope and desire of the editors that you will "get the most out of your drill press," after reading this book.

Power Tool Division

Copyright © 1937, 1952, 1954, Rockwell Manufacturing Company
Copyright © 1978 Rockwell International Corporation
Published by Rockwell International

Brief quotations may be used in critical articles and reviews. For any other reproduction of this book, including electronic, mechanical, photocopying, recording or other means, written permission must be obtained from the publisher.

Text prepared and book designed by
Robert Scharff & Associates

Library of Congress Catalog Card Number: 78-56312

Manufactured in the United States of America

Contents

FORWARD	iv
1. GETTING TO KNOW THE DRILL PRESS	1
2. BEFORE OPERATING THE DRILL PRESS	25
3. BORING IN WOOD	45
4. MORTISING ON THE DRILL PRESS	71
5. DRILLING IN METAL	81
6. SANDING ON THE DRILL PRESS	97
7. OTHER OPERATIONS ON THE DRILL PRESS	107
8. SHARPENING DRILLS AND BITS	111
INCH/MILLIMETER CONVERSIONS	119
INDEX	120

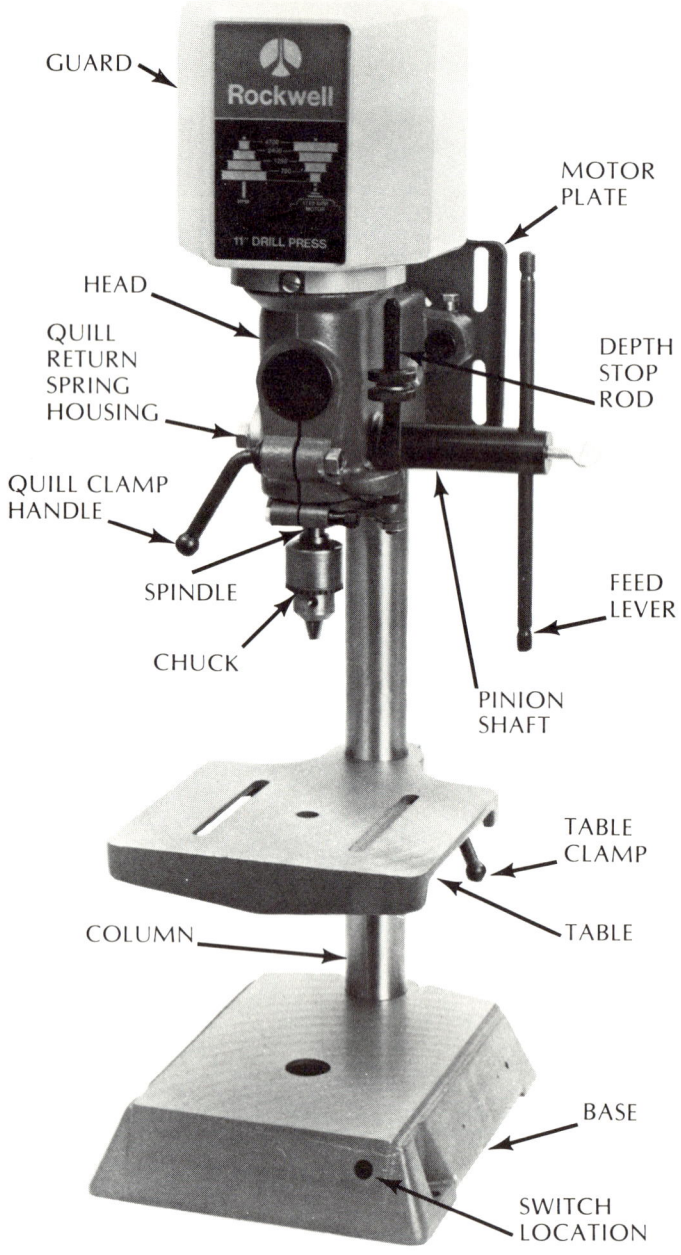

Fig. 1-1: Typical 11-inch bench model drill press with parts.

Chapter 1

GETTING TO KNOW THE DRILL PRESS

The drill press was designed originally for the metalworking trades. However, with the availability of woodworking techniques and of cutting tools, jigs, and attachments, the drill press is now one of the most versatile tools in the shop. It not only drills in metal, but also bores in wood and performs other woodworking operations such as mortising and sanding. In fact, after the table saw, the drill press can easily become the second most important piece of equipment in the average home workshop.

CONSTRUCTION AND SIZES

The conventional drill press (Fig. 1-1) consists of the following main parts: the base, column, table, and head. The base supports the machine. Usually, it has three drilled holes for fastening the drill press to the floor or a stand or bench.

The column, generally made of steel, holds the table and head and is fastened to the base. Actually the length of this hollow column determines whether the drill press is a bench model (Fig. 1-2A) or a floor model (Fig. 1-2B). That is, floor machines range in height from 66 to 72 inches, bench models, from 36 to 44 inches.

The table is clamped to the column and can be moved to any point between the head and the base. The table may have slots in it to aid in clamping holding fixtures or the workpiece. It usually also has a central hole through it. Some tables can be tilted to any angle, right or left, while other models have a fixed position only. An auxiliary table made of plywood or particleboard (chip board), which can be readily fastened to the regular drill press table,

Fig. 1-2: (A) an 11-inch bench model drill press mounted on a metal stand; (B) a 15-inch floor model drill press.

1

is available on some models, or it can be made in the shop.

The term *head* is used to designate the entire working mechanism attached to the upper part of the column. The essential part of the head is the spindle. This revolves in a vertical position and is housed in bearings at either end of a movable sleeve, called the quill. The quill, and hence the spindle which it carries, is moved downward by means of a simple rack-and-pinion gearing, worked by the feed lever. When the feed handle is released, the quill is returned to its normal up position by means of a spring. Adjustments are provided for locking the quill and presetting the depth to which the quill can travel. Incidentally, the quill usually has a stroke or travel of from 3 to 4 inches in most home workshop models (Fig. 1-3).

Fig. 1-3: How quill travel is measured. The drill press illustrated here has a quill travel of 3 5/8 inches.

To increase the versatility of the drill press, some manufacturers offer interchangeable spindles. The standard spindle for most drill presses has a 1/2-inch capacity geared chuck with a key. This chuck (Fig. 1-4) offers the best grip for most work. Another type of interchangeable spindle is shown on page 15. Most drill press accessories fit directly into the geared chuck (Fig. 1-5).

The spindle is usually driven by a stepped cone pulley connected by a V belt to a similar pulley on a motor, which is usually bolted to a plate on the head casting in the rear of the column. The average range of

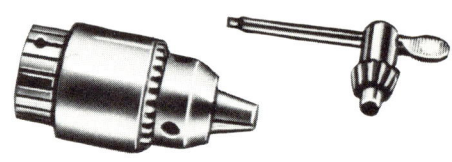

Fig. 1-4: Standard geared chuck and key.

speeds is from 600, to about 5000 revolutons per minute (rpm). When the machine is used exclusively for metal work, a larger cone pulley is used on the spindle to give speeds of about 450, to 2000 rpm. Because the motor shaft stands vertically, a sealed ball-bearing motor should be used as a power unit. For average work, a 1/3- or 1/2-horsepower motor meets most requirements.

The capacity or size of the drill press is determined by the distance from the center of the chuck to the column. This distance is expressed as a diameter. For example, an 11-inch drill press will drill a hole through the center of a round piece 11 inches in diameter. The actual distance from the center of the chuck to the front of the column is 5 1/2 inches. Conventional drill press sizes for home workshops generally range from 11 to 15 inches.

Radial Drill Press. The radial or universal type drill press (Fig. 1-6) is a highly versatile

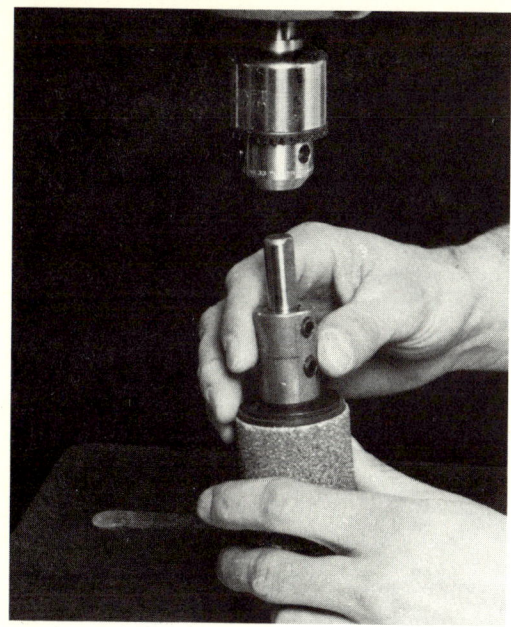

Fig. 1-5: Many accessories have special adapters to fit the geared chuck.

tool. It does jobs that may be impossible on conventional-type drill presses. The head swivels 360 degrees around the column and tilts more than 90 degrees right and left. The ram in the model shown travels 11 inches back and forth. You can drill to the center of a 32-inch circle, can do horizontal and angular drilling, multiple or series hole drilling, and more. The operations of both the radial drill press and the conventional type are discussed in this book.

WHAT TO LOOK FOR WHEN BUYING A DRILL PRESS

As already stated, the drill press is one of the most versatile of power tools. For that reason, it is essential that it should have qualities and operating features that permit it to be utilized fully. Listed below are a number of features to look for when buying a drill press for the home workshop.

1. The entire drill press should be solidly constructed to allow for long life and continuing precision work.
2. The table and base (Fig. 1-7) should be ribbed for strength and rigidity. They should be slotted. The table should have flats or ledges on the sides, used for clamping the the work. (This offers convenience and safety to the user.) The table should be ground for the accurate work and the base should also have a flat surface for holding large workpieces.
3. The head should be cast iron since it offers excellent support and protection for the most important parts of the drill press which are the motor, quill and pinion shaft.
4. The drill press should be equipped with a chuck that is tightened by a wrench or key rather than by hand. The chuck should have a 1/2-inch capacity so that it will accommodate the various size bits and accessories. Many drills feature a taper-mounted chuck. By having a taper-mounted chuck, the runout is practically eliminated and the user is assured of accurate drilling. Some chucks feature a self-ejecting key which insures that the key is not left accidentally in the chuck.
5. The depth-adjustment gauge allows the user to drill many holes at the same depth as the original hole. It eliminates any

Fig. 1-6: Radial type drill press.

3

guessing and allows precision accuracy.

6. Proper provision should be made for adjustment in case of quill wear. This will help to insure the lasting accuracy of the drill press.

7. The drill press should have an adjustable motor bracket support, sturdily constructed to support the motor, yet easily movable to assure proper belt tension.

8. The drill press should have a selection of speeds for drilling wood, metal, plastic, glass, and ceramics. Most drills feature a four-speed pulley (Fig. 1-8) to accomplish multiple speeds.

9. Be certain you can get proper replacement parts and service, if needed.

10. A complete line of accessories will help you get the most from your machine. Accessories which are supplied by the manufacturer of the tool are designed for the particular tool that you buy, and it will not be necessary to use makeshift arrangements to use them to their best advantage.

11. It costs very little more at the start, and much less in the long run to equip your workshop with the best in power tools. Choose a drill press produced by a manufacturer who has established a record of quality and reliability.

Fig. 1-7: A cast iron base and table give the necessary support and stability for heavy drilling applications.

BITS AND DRILLS

To be technically correct, you "bore" a hole in wood (even though twist drills are used) and "drill" a hole in metal. Today, however, this terminology is not followed rigidly, and usually the two terms—bore and drill—are used interchangeably in discussion of power drills. However, different bits are usually used to cut in wood and in metals.

Wood Bits. Several types of wood bits are used for wood boring with a drill press. Figure 1-9 shows the two common styles of wood bits. One has a screw point, the other a brad or diamond point. Figure 1-9A has a solid center, with a single spiral running around it; Fig. 1-9B is a double-twist bit. The cutting edges are very similar. The opening in the spiral is called the *throat*. In some styles of double-twist bits it is designated by the term *flute*. Both terms mean the same thing.

Brad points are recommended for use in most drill press work done in the home shop. When screw-point bits are used on the drill press, the rate of feed must be the same as the lead of the screw; otherwise the bit will lift the work and prove generally unsatisfactory. Where screw-point bits are a part of the shop equipment, it is advisable to partly file off the threads so that the point acts the same as a brad point.

Figure 1-10A shows a *hollow-spiral bit*. It has a screw point and one cutting edge, and differs from all others in that the center is hollow so as to permit the passage of chips. This bit cuts very fast and is used extensively for deep holes. It does not make a smooth hole. Figure 1-10B is a *double-spur bit*. This term is quite general and describes the fact that the bit has two spurs at the end. Figure 1-10B could also be described as a double-twist fluted bit, while Fig. 1-10C is a *single-twist solid-center bit*. While the solid-center bit has two cutting edges, one of these terminates directly in back of the head. Both of these bits cut rapidly and smoothly. The drill shown in Fig. 1-10B

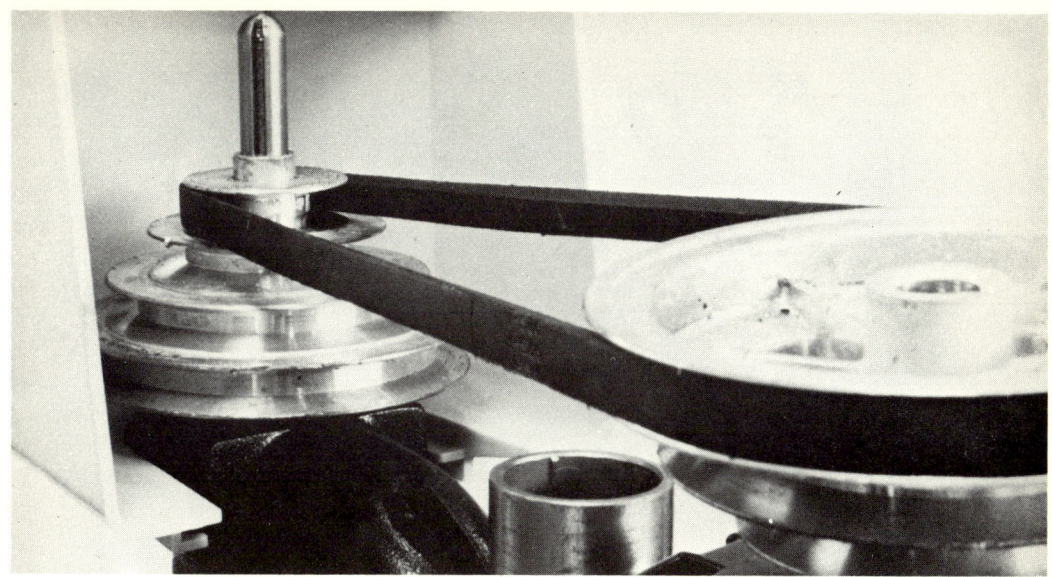

Fig. 1-8: A typical four-speed pulley arrangement.

gives the cleaner hole; the drill in Fig. 1-10C is the stiffest.

To drill larger holes in wood the *spade or speed-type* bit (Fig. 1-11) is generally recommended. Bits of this style range from about 3/8 inch to as large as 1 1/2 inches. They relieve the chips easily, and binding is not much of a problem. Spade or speed bits have a tendency to split not only the front surface of the wood, but the back surface as well. The front surface can be drilled without splintering by starting the hole slowly. (That is, do not press too hard. Also, be sure the bit goes in square to the wood.) The back surface can be drilled clean without splintering by either of two methods. First, as soon as the pilot (the center of the bit) comes through the back, stop drilling. Complete the hole by drilling from the back. Or, second, place a piece of scrap behind or under the workpieces, and drill through the workpiece into the scrap piece.

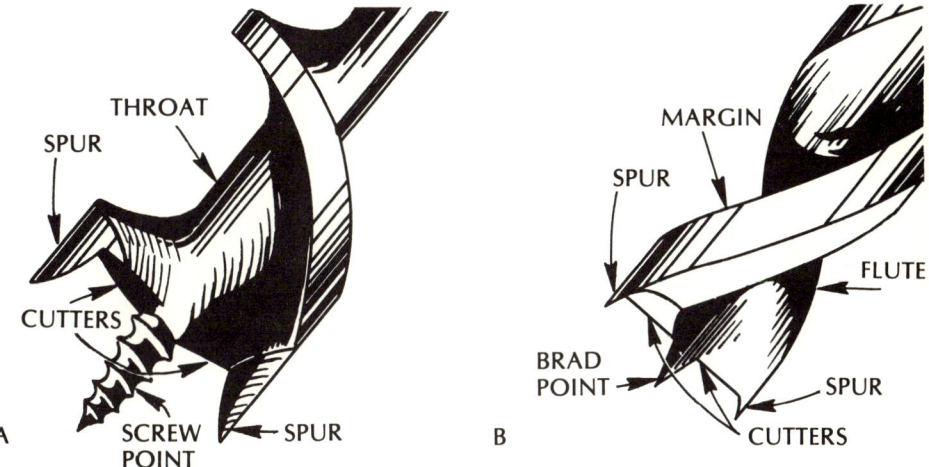

Fig. 1-9: (A) A solid center bit with a screwpoint, (B) machine spur bit with a bradpoint.

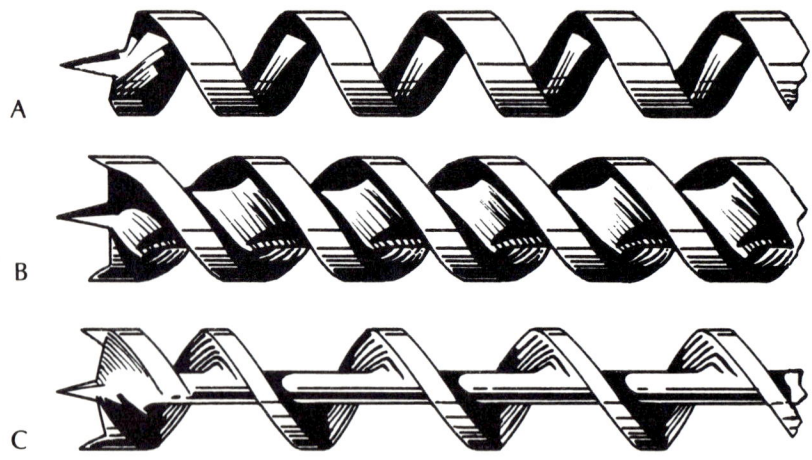

Fig. 1-10: Three common auger wood boring bits: (A) Hollow-spiral bit; (B) double-spur bit and (C) single-twist solid-center bit.

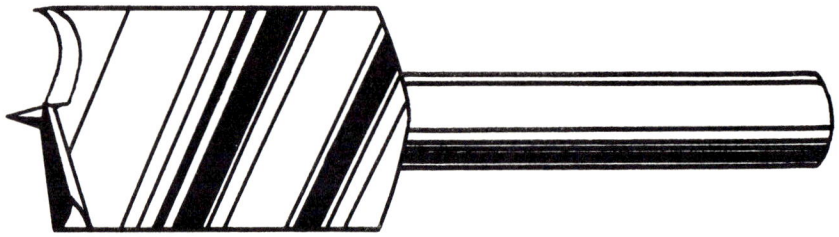

Fig. 1-11: Spade or speed-type bit.

Holes larger than 1 1/2 inches can be cut with a *rotary hole saw* (Fig. 1-12). This is literally a saw bent into a circle. It will make a clean, round hole in anything a hacksaw will cut, including metal, plastic, composition board, and asbestos. The pilot bit of the hole saw mandrel can be centered on a punch mark to locate the large hole with great accuracy. Although hole saws are available in a size range from about 3/4 to 6 inches in diameter, most home workshop drill presses are not powerful enough to cut holes larger than 4 inches in diameter with hole saws.

One-size rotary hole saws are more expensive. More economical is the type with a shaft or mandrel on which saws of various sizes can be mounted. The cuplike saw shells range from approximately 5/8 to 2 1/2 inches in diameter (Fig. 1-13) and are deep enough to cut through 3/4-inch-thick material. (A few are designed to cut up to 2 inches.) Be sure to tighten the chuck for maximum grip when using a hole saw, since its large diameter puts great stress on the spindle. At any angle other than 90 degrees the hole saw will start cutting on one edge instead of all around, so take pains to start the pilot drill straight. In thick or hard wood, withdraw the saw occasionally to clear the chips and help cool the hole saw.

Fig. 1-12: The mandrel and four different sizes of rotary hole saws.

Fly-cutter-type circle makers (Fig. 1-14) are also available for cutting holes from 1/2 to 8 inches in diameter. Actually, the size of the hole is controlled by loosening a setscrew and then sliding the cutter blade in or out. For best results when cutting circles in wood, cut halfway through each side of the wood, or back up the wood, so that the cutting blade does not tear and splinter the wood as it comes through. The cutter blade should be set back behind the center drill bit approximately 1/2 inch (where the flutes end on the drill bit), so that the blade will be held firmly in place when it begins to bite into the wood. Since the circle cutter has an off-center load, it works more smoothly and with less vibration at slower speeds. When using a fly cutter, be sure that the workpiece is securely clamped to a solid surface.

With either the rotary hole saw or fly cutter, splintering of the far side of the work will be prevented if you bore the hole about halfway through and then finish cutting it from the other face of the work. The pilot hole, having passed through the work, centers the tool for its second cut.

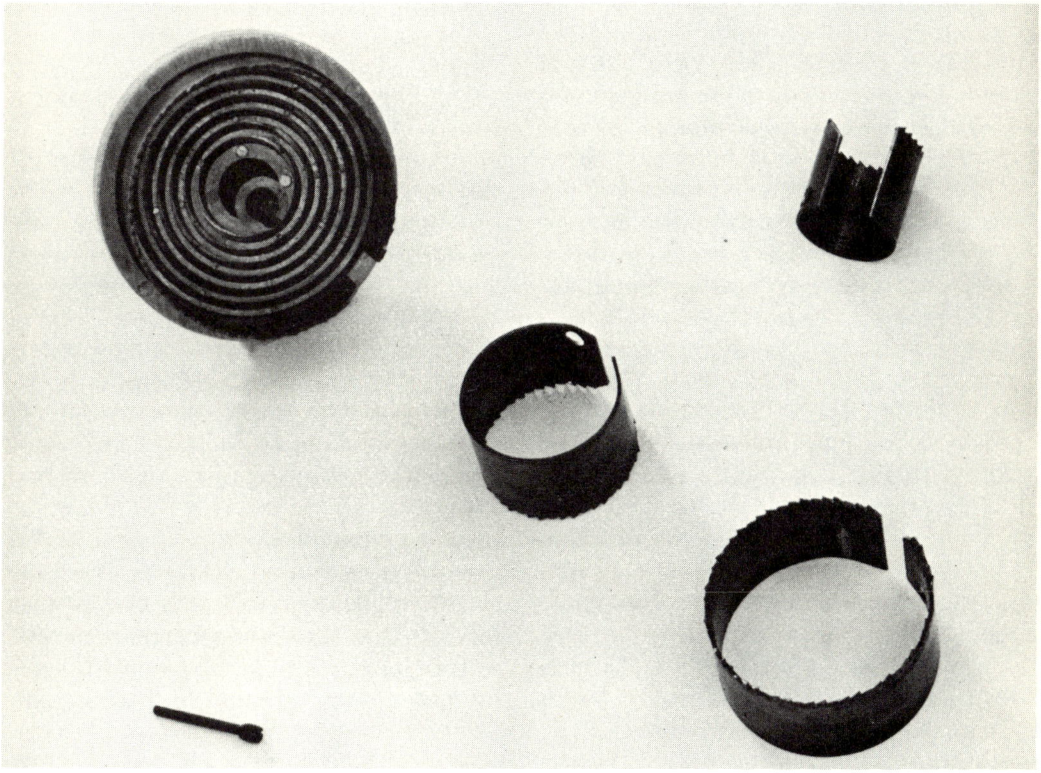

Fig. 1-13 A popular 4-in-1 hole saw.

There are other types of bits for drilling holes in wood, such as the *spur, auger, self-feed, expansive,* and *Foerstner* bits.

The *machine spur* bit (Fig. 1-15A) has a brad and lip point and is one of the cleanest, fastest-cutting bits for dowel holes. These bits come in standard sizes from 1/4 to 1 inch and are generally available in thirty-seconds of an inch. The *multispur bits* (Fig. 1-15B) give best results for larger holes, from 1/2 to 1 1/2 inches.

Power auger bits for drill presses have a straight shank and a brad point; they never have a square tang (which is only used in a hand-operated brace—page 14). They come in sizes as small as 1/4 inch and as large as 2 inches. (Those larger than 1 inch are primarily for industrial use.) Common power auger bits are made either with solid-center or double-spur bits (Fig. 1-10B and C). Because of their efficient chip ejection, auger bits are best for drilling deep holes. The spurs on the auger score the circumference of the hole and then the lips cut out the shavings as they revolve. These bits bite deeply into the wood, so the work should be held securely with clamps to prevent movement. Boring the hole from both sides or clamping a block of wood to the back of the work will prevent the bit from splintering the wood as it breaks through. Never use an ordinary screw-point auger in a drill press. The point will screw itself into the wood and jam there, stalling the motor. Such screw-point augers are meant for turning by hand, with the screw helping to feed the bit into the work.

Self-feed bits (Fig. 1-15C) are used for drilling large-diameter holes in wood. They have a specially-designed point that draws the bit into the wood. For this reason they should only be used in drills with slow speeds.

The *Foerstner bit* Fig. 1-15D) is a cutting tool used for drilling *blind holes,* (holes that do not go all the way through the workpiece). It leaves the bottom of the hole flat. It can be used to drill completely

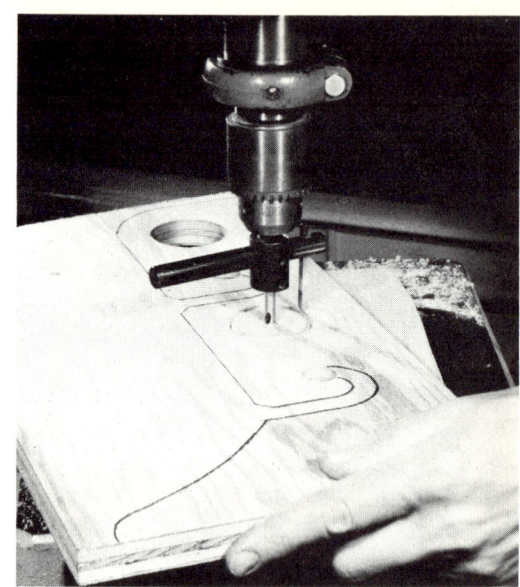

Fig. 1-14: Typical fly-cutter-type circle maker.

through the workpiece or for enlarging previously drilled holes, and for other special applications. For drill press use, they come in all sizes, up to 4 inches in diameter.

The common style of *expansive bit* is shown in Fig. 1-15E. The single cutter can be moved in or out in the groove to cut any size hole within its capacity. This bit can be used on the drill press, but the work must be securely clamped because the single cutter is an off-center load. It should also be run at lower speeds.

It is a good idea, before driving screws in wood, to drill a pilot hole. Not only is it much easier to drive a screw in a workpiece with a pilot hole (especially with a hand screwdriver), but the pilot hole also prevents splitting of the wood. There is a correct size (diameter) of pilot hole for every size of screw. For hardwoods, pilot holes are slightly larger and a little deeper than for soft-woods, because hardwoods tend to split more easily. If you use twist bits to drill pilot holes, consult the chart on page 51, for the size of the bit to be used to drill the correct hole for each size of screw. There are also special profile bits, made for

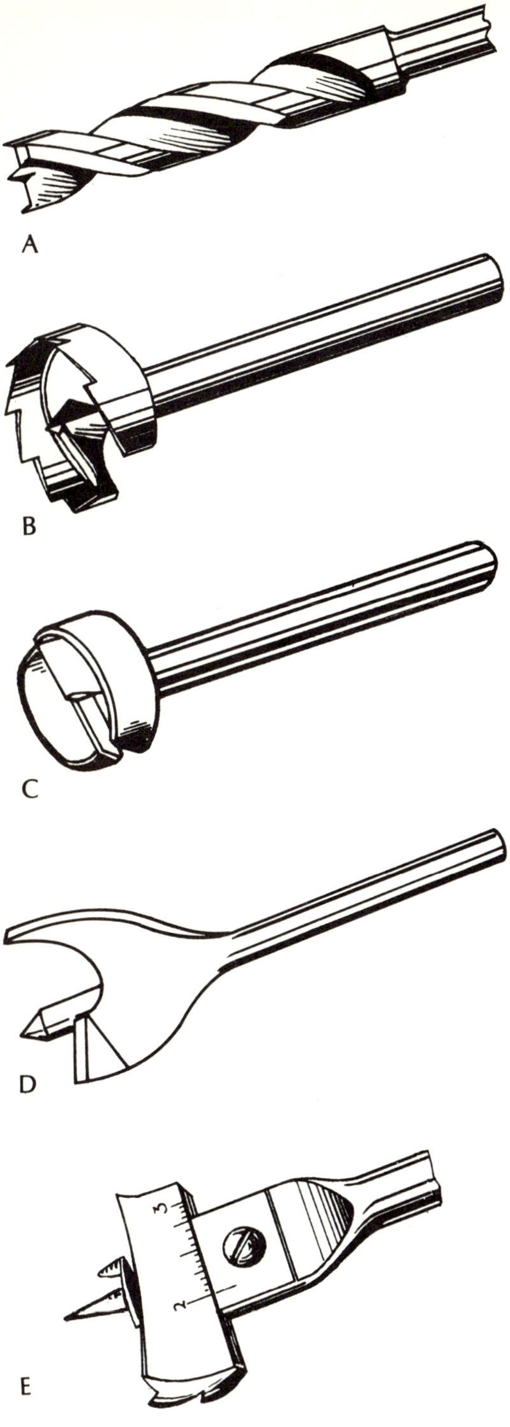

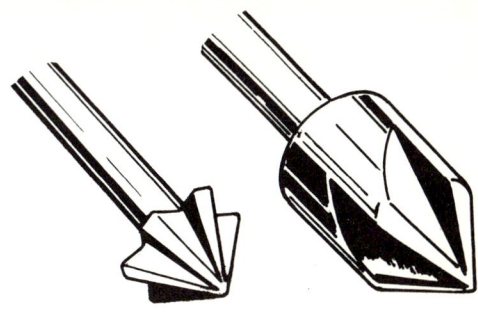

Fig. 1-16: Two types of countersink bits.

drilling pilot holes. Each is stamped with the number or size of the screw for which they are intended. Most will also bevel the hole for the flush fit of flathead screws (Fig. 1-16). There are also a few which will drill, countersink and counterbore the holes for wood plugs to conceal the screwheads (Fig. 1-17)

Plug cutters are available as attachments for drill presses (Fig. 1-18). They can be obtained in sizes from 3/8 to 1 inch, by sixteenths. Cross-grained plugs up to 1-inch thick can be cleared through the center opening, while the full length of the cutter will make dowels up to 2 inches long.

One of the essentials in keeping wood bits in good condition is to keep them well polished. When the flutes are clean and smooth, the chips are ejected easily, whereas a rusty bit will not throw out the chips and will clog and burn in deep holes. Bits are easily kept clean by lightly rubbing them with steel wool, as shown in Fig. 1-19. A thin film of oil should be placed on bits used infrequently.

Wood bits are sharpened as described in Chapter 8.

Twist Drills. While twist drills are designed for metal drilling, they can be used to make holes in most materials—wood, plastic, ceramic, or other materials. They are often rated as the most efficient of all cutting tools, based on the length of the cutting edge in proportion to the amount of metal which supports it. It is a tool which

Fig. 1-15: Other popular wood boring bits: (A) spur machine bit; (B) multispur bit; (C) self-feed bit; (D) Foerstner bit; and (E) expansive bit.

Fig. 1-17: A screw drill that drills holes, countersinks and counterbores in one operation. It can be adjusted for any screw length, and for flat, round or oval head screws.

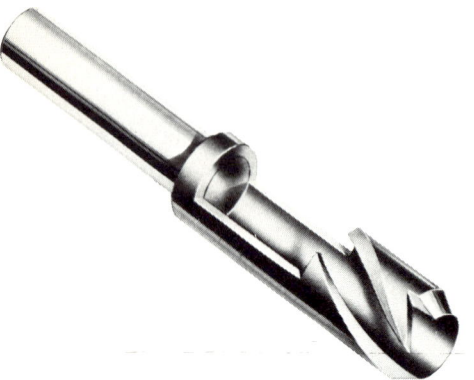

Fig. 1-18: A typical plug cutter.

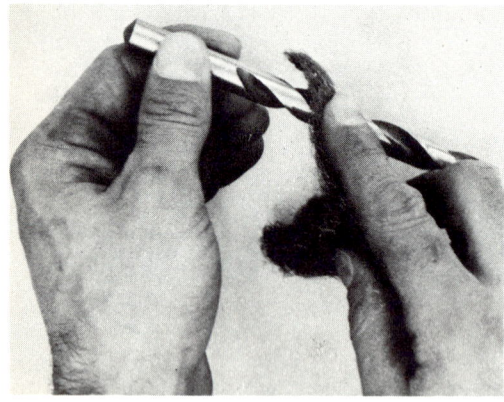

Fig. 1-19: Bits should be kept polished so that they will eject the chips.

will stand a tremendous amount of abuse and still keep cutting. Nevertheless, any drill will work better and last longer if properly ground. Grinding is an operation that can be done after a brief study of the fundamental mechanics of a good cutting edge (see Chapter 8).

The principal parts of a twist drill (Fig. 1-20) are the body, the shank, and the point. The dead center of a drill is the sharp edge at the extreme tip end of the drill. It is formed by the intersection of the cone-shaped surfaces of the point and should always be equally divided by the axis of the drill. The point of the drill should not be confused with the dead center. The point is the entire cone-shaped surface at the end of the drill.

The lip or cutting edge of a drill is that part of the point that actually cuts away the metal when a hole is drilled. It is ordinarily as sharp as the edge of a knife. The point angle has been established as 118 degrees for general work, and this angle should be maintained. However, for extensive drilling in wood, a much sharper angle should be used. A standard 118 degree angle drill is easily checked with a drill-point gauge. Gauges in a variety of styles can be purchased at a nominal cost, or can be made

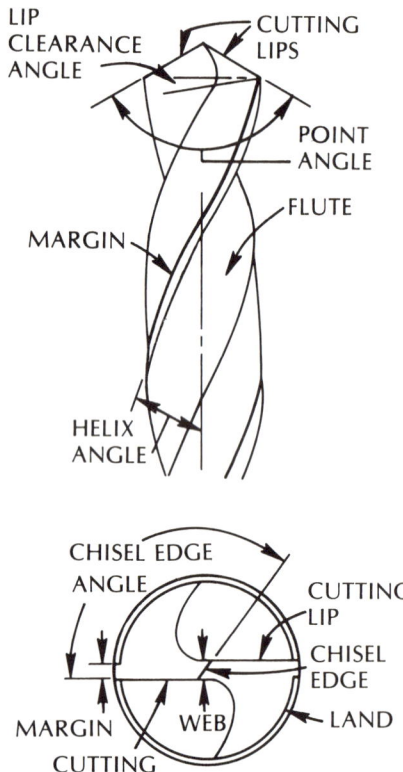

Fig. 1-20: Twist drill nomenclature.

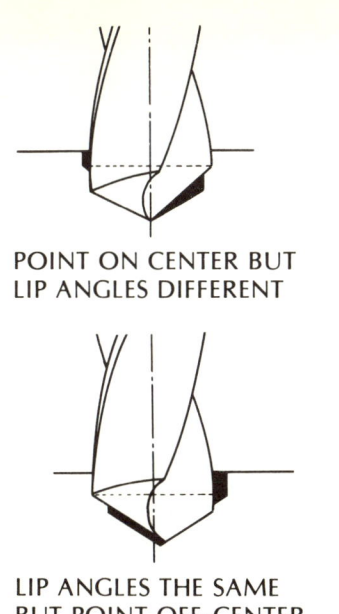

POINT ON CENTER BUT
LIP ANGLES DIFFERENT

LIP ANGLES THE SAME
BUT POINT OFF-CENTER

Fig. 1-21: The lip angle of a twist drill is an important feature.

from sheet metal. The markings on the edge need not be exact since they are used only to check the length of one lip against the other. In use, the drill body is held against the edge of the gauge, and in such a position that the angular edge is over the cutting lip of the drill. The gauge will then show whether or not the point, or, rather, one edge of the point, is at the correct angle. Besides being ground to the correct angle, both lips must be exactly the same length. What happens when the lip angles are different or of unequal length is shown in Fig. 1-21. It can be seen that the resulting hole will be out of round and larger than the drill.

There is a cutting edge for each flute of the drill. Like any other cutting tool, there must be clearance behind the cutting edge before the drill can cut. This clearance can be seen readily on a properly-ground drill by using the drill-point gauge, placing it

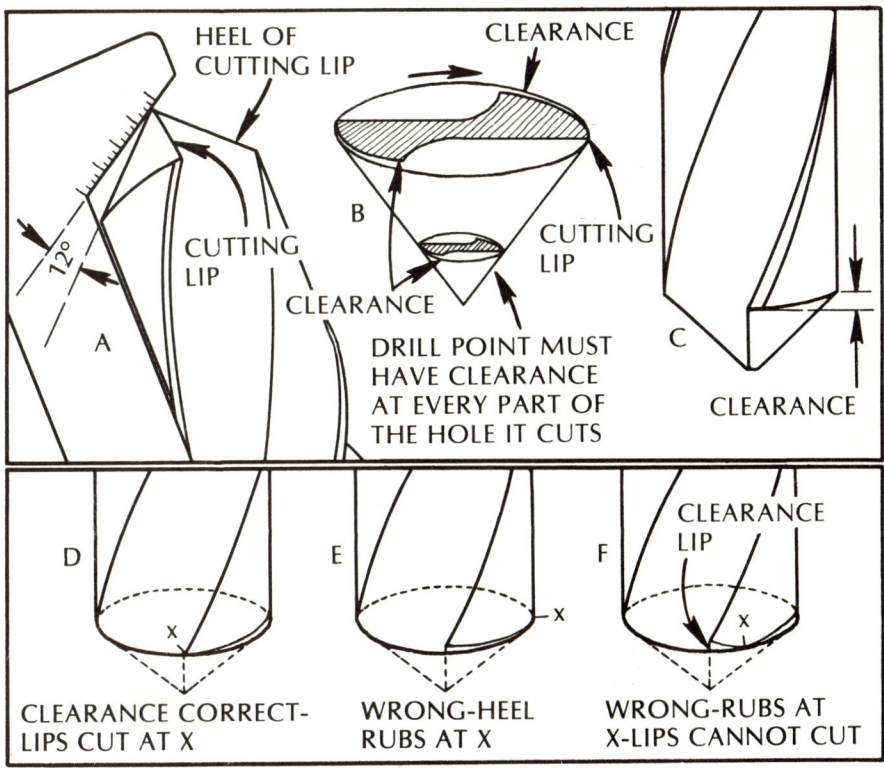

Fig. 1-22: Proper lip clearance is an important feature of a twist drill.

11

DECIMAL EQUIVALENTS OF DRILL SIZES

Drill	Decimal Diameter In Inches	Decimal Diameter In Millimeters	Drill	Decimal Diameter In Inches	Decimal Diameter In Millimeters
80	0.0135	0.3429	46	0.081	2.0574
79	0.0145	0.3683	45	0.082	2.0828
1/64	0.0156	0.3969	44	0.086	2.1844
78	0.016	0.4064	43	0.089	2.2606
77	0.018	0.4572	42	0.0935	2.3749
76	0.02	0.508	3/32	0.09375	2.3812
75	0.021	0.5334	41	0.096	2.4384
74	0.0225	0.5715	40	0.098	2.4892
73	0.024	0.6096	39	0.0995	2.5273
72	0.025	0.635	38	0.1015	2.5781
71	0.026	0.6604	37	0.104	2.6416
70	0.028	0.7112	36	0.1055	2.6797
69	0.0292	0.7417	7/64	0.109375	2.7781
68	0.031	0.7814	35	0.11	2.794
1/32	0.03125	0.79375	34	0.111	2.8194
67	0.032	0.8128	33	0.113	2.8702
66	0.033	0.8382	32	0.116	2.9464
65	0.035	0.889	31	0.12	3.048
64	0.036	0.9144	1/8	0.125	3.175
63	0.037	0.9398	30	0.1285	3.2639
62	0.038	0.9652	29	0.136	3.4544
61	0.039	0.9906	28	0.1405	3.5687
60	0.04	1.016	9/64	0.140625	3.5719
59	0.041	1.0414	27	0.144	3.6576
58	0.042	1.0668	26	0.147	3.7338
57	0.043	1.0922	25	0.1495	3.7973
56	0.0465	1.1811	24	0.152	3.8608
3/64	0.046875	1.191	23	0.154	3.9116
55	0.052	1.3208	5/32	0.15625	3.9688
54	0.055	1.397	22	0.157	3.9878
53	0.0595	1.511	21	0.159	4.0386
1/16	0.0625	1.5875	20	0.161	4.0894
52	0.0635	1.6129	19	0.166	4.2164
51	0.067	1.7018	18	0.1695	4.3053
50	0.07	1.778	11/64	0.171875	4.3656
49	0.073	1.8542	17	0.173	4.3942
48	0.076	1.9304	16	0.177	4.4958
5/64	0.078125	1.9844	15	0.18	4.572
47	0.0785	1.9939	14	0.182	4.6228

Drill	Decimal Diameter In Inches	Decimal Diameter In Millimeters	Drill	Decimal Diameter In Inches	Decimal Diameter In Millimeters
13	0.185	4.699	Q	0.332	8.4328
3/16	0.1875	4.7625	R	0.339	8.6106
12	0.189	4.8006	11/32	0.34375	8.7312
11	0.191	4.8514	S	0.348	8.8392
10	0.1935	4.9149	T	0.358	9.0932
9	0.196	4.9784	23/64	0.359375	9.1281
8	0.199	5.0546	U	0.368	9.3472
7	0.201	5.1054	3/8	0.375	9.525
13/64	0.203125	5.1594	V	0.377	9.5758
6	0.204	5.1816	W	0.386	9.8044
5	0.2055	5.2197	25/64	0.390625	9.9219
4	0.209	5.3086	X	0.397	10.0838
3	0.213	5.4102	Y	0.404	10.2616
7/32	0.21875	5.5562	13/32	0.40625	10.3188
2	0.221	5.6134	Z	0.413	10.4902
1	0.228	5.7912	27/64	0.421875	10.7156
A	0.234	5.9436	7/16	0.4375	11.1125
15/64	0.234375	5.9531	29/64	0.453125	11.5094
B	0.238	6.0452	15/32	0.46875	11.9062
C	0.242	6.1468	31/64	0.484375	12.3031
D	0.246	6.2484	1/2	0.5	12.7
E	0.25	6.35			
1/4	0.25	6.35			
F	0.257	6.5278			
G	0.261	6.6294			
17/64	0.265625	6.7469			
H	0.266	6.7564			
I	0.272	6.9088			
J	0.277	7.0358			
K	0.281	7.1374			
9/32	0.28125	7.1438			
L	0.29	7.366			
M	0.295	7.493			
19/64	0.296875	7.5406			
N	0.302	7.6708			
5/16	0.3125	7.9375			
O	0.316	8.0264			
P	0.323	8.2042			
21/64	0.328125	8.3344			

NOTE: Commercial gauges are available but this homemade one is useful.

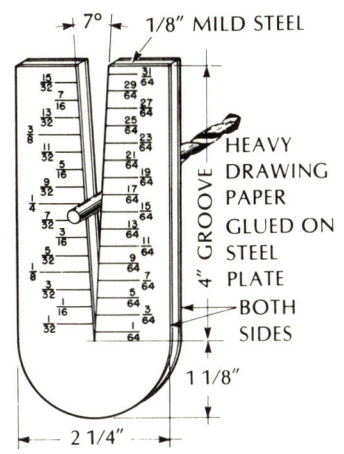

over the heel of the point, as shown in Fig. 1-22A. It will be noted that the angle here is 12 degrees less than the lip angle, and this is within the proper clearance for the average drill—from 12 to 15 degrees. Clearance can also be observed by holding the drill as in Fig. 1-22C and noting the difference between the lip and heel of the point. Two horizontal sections of a drill point are shown in Fig. 1-22B. It can be seen that there must be clearance behind the cutting lips at every part of the conic recess which the drill cuts. With proper clearance (Fig. 1-22D), the drill cuts at the cutting lips, leaving every part of the point behind the lips in the clear. Figure 1-22E shows just the reverse of correct clearance—the drill rubs at the heel and the lips cannot cut. Likewise, if any part of the heel rubs against the conic recess (Fig. 1-22F), the lips cannot cut.

The strip along the inner edge of the body is called the *margin*. It is the greatest diameter of the drill and extends the entire length of the flute. The diameter of the margin at the shank end of the drill is smaller than the diameter at the point. This allows the drill to revolve without binding when deep holes are drilled. Since the twist drill has no spur or screw at its end to pull it into the work, pressure must be applied to do its cutting. It cuts away the material at the bottom of the hole as the drill is pushed into the work while it turns. Attention must be given to the condition of the tip of the drill. Dull and broken tips heat very rapidly and will not cut straight, clean holes.

The shank (Fig. 1-23) is the part of the drill which fits into the spindle or chuck of the drill press. As mentioned earlier, drills used on the drill press commonly have straight shanks to fit the adjustable chuck on the machine. It is also possible to mount taper shank drills by using a taper shank spindle.

Twist drills are made of either carbon-steel or high-speed steel. High-speed steel drills, made of an alloy which usually contains tungsten, chromium and vanadium, are designed expressly for work on metal and can take considerable heat without weakening or becoming dull. Usually, high-speed drills can do their work without the use of a coolant. The carbon-steel drills are softer and are used solely on wood and soft metals or plastics. They cost much less than the harder high-speed steel drills but will wear quickly and become distorted if overworked. On soft metals, they require a flow of cooling liquid on the tip, to prevent burning (see Chapter 5). Carbide-tipped drills are also available. These are primarily used for drilling masonry, ceramics, and extra-hard materials.

Twist-drill sizes are denoted by three different systems. The table here shows that the smallest drills are numbered by wire-gauge sizes from 1 to 80, the largest being number 1, which is 0.228 inch in diameter. Number 80 measures 0.0135 inch in diameter. Letter-size twist drills range from size A, which is 0.234 inch, to Z, which is 0.413 inch in diameter. The third series of twist drills overlaps the other two, but without duplication. These are denoted in fractional sizes, increasing by sixty-fourths of an inch from 1/32 to 1/2 inch. Straight-shank and straight fluted drills for wood and soft metal are sold in sets of 16 sizes from 1/32 to 1/2 inch. It is usually necessary to have a complete set of twist drills on the workbench. For the average home shop, a set of fractional-size drills

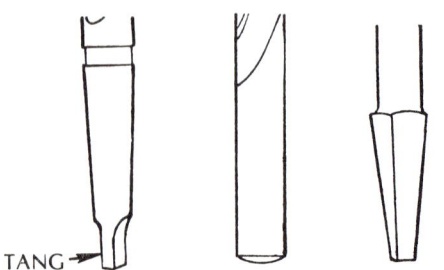

Fig. 1-23: Representative shanks: (Left) Tapered, (center) straight, and (right) square tang. The latter is used primarily for hand-operated braces—not drill presses (see page 8).

Fig. 1-24: Mortising equipment installed on a drill press.

Fig. 1-25: Mortising chisel and chisel bit.

does very well. They should be kept in a drill stand, however, so that the correct size can be selected readily. Inexpensive drill stands are available with the size marked alongside each hole. Metric-sized twist drills are also available.

Mortising Bits and Chisels. Mortising equipment (Fig. 1-24) includes a hollow chisel with four cutting edges that cut a square hole. The bit is similar to an ordinary wood bit without a point, which works inside the chisel and removes the bulk of the wood. Even without a point, the bit makes a straight cut, because it is supported by the chisel. It is desirable not to have a point, so the bottom of the mortise will be flat. Bushings are usually supplied for the bits so that any size bit can be mounted in the 1/2-inch hole spindle. In some cases the bit can be directly mounted in the geared chuck.

Mortising bits and chisels for use with the mortising attachment (Fig. 1-25) are generally available in 1/4, 5/16, 3/8, or 1/2 inch sizes, with the 3/8-inch size being used the most. They are made of select steel and will hold their cutting edges.

Accurately ground bushings are used when mounting bits in a 1/2 inch hole spindle.

DRILL PRESS ACCESSORIES

Proper accessories enable you to do more jobs and do them easier and faster. Some of the more common accessories available for most drill presses are discussed here.

Interchangeable Spindles. As mentioned previously, interchangeable spindles allow selection of the right spindle for the job and increase the versatility of the machine many times. They also add to the life of the machine.. Work can be kept close to the bearings on some operations. This also makes it easier and faster to change cutting tools, since the entire spindle can be removed and replaced readily. Generally, two different spindles (Fig. 1-26) are available.

The *geared chuck spindle*, 1/2-inch capacity, is used for general drill press work. This spindle is usually the type that comes with the drill press.

The *1/2-inch hole spindle* is used for all cutting tools with 1/2-inch diameter shanks. It can also be used with the mortising attachments, drum sanders, and some other accessories.

Mortising Attachment. The fence, hold-down, hooked rods, and chisel holder shown in Fig. 1-24 make up the attachment that converts a drill press into an accurate

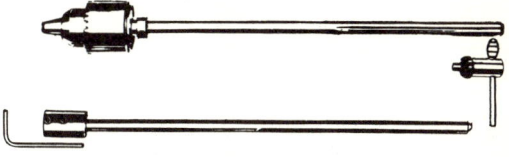

Fig. 1-26: The two most common spindles.

15

mortising tool that makes straight, true, square-end mortises of any length in a fraction of the time needed, using hand methods. Capacity under the hold-down is about 4 3/4-inches.

The *foot feed* (Fig. 1-27), although not absolutely necessary, is recommended for use with the mortising attachment. It frees both hands of the operator or, when used in connection with the regular hand feed lever, permits the rapid penetration needed for making clean mortises without burning the chisel or bit. The foot feed is suitable only for the floor model machines.

Sanding Drums and Abrasive Sleeves
Sanding on curved surfaces can be accomplished on the drill press with the use of a sanding drum (Fig. 1-28). Various sizes of sanding drums are available. Aluminum oxide-coated sleeves for metal and garnet-coated sanding sleeves for wood are made up in different grits and sizes to fit the various drums.

The 11/16-inch wide drum shown at the bottom of the illustration is held in the jaws of the geared drill chuck. The rest of the drums shown are best held in the accessory 1/2-inch hole spindle. The small drum has a 1/2-inch shank to fit the 1/2-inch hole spindle; larger sizes have a 1/2-inch hole and are fitted to the spindle by means of a short length of 1/2-inch metal rod (see Fig. 1-5).

Disk sanding can also be done on the

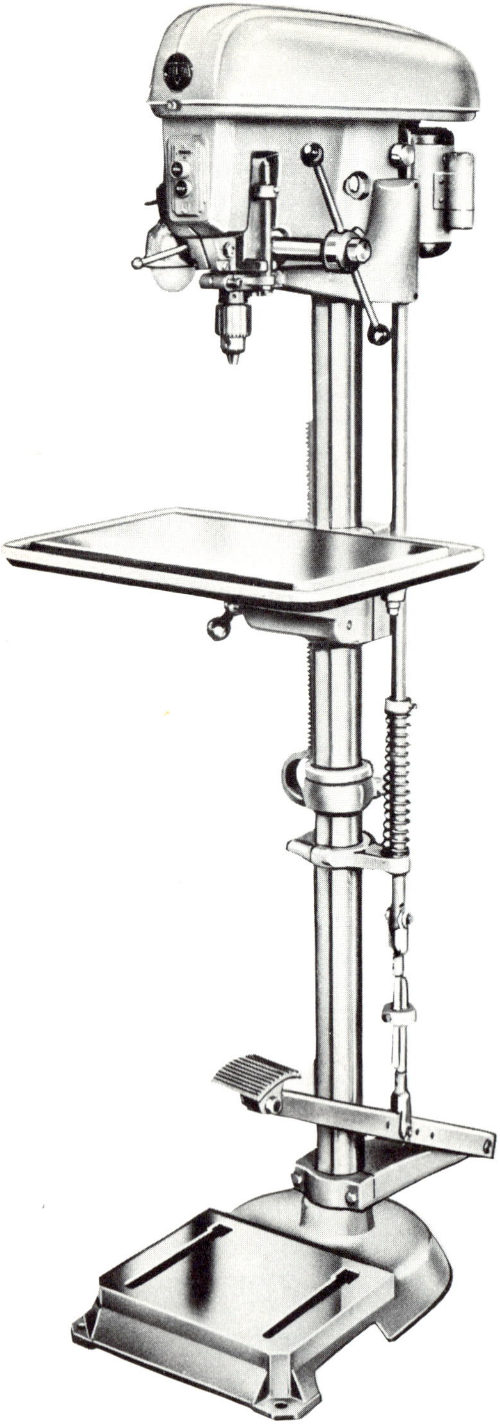

Fig. 1-27: Foot feet is suitable for floor type drill presses that have column height of about 64 inches.

Fig. 1-28: Three popular sanding drum sizes: (Left to right): 3-inch, 2-inch, and 11/16-inch.

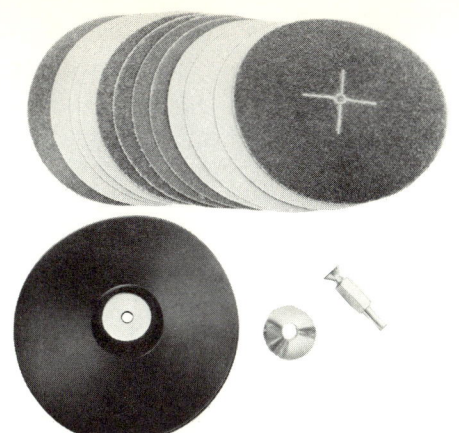

Fig. 1-29: Items that make up a typical sanding kit: (Top) various grits and types of sandpaper disks: (bottom) rubber backing pad; washer and 1/4-inch shank arbor adapter.

drill press. To accomplish this, a sanding disk kit such as the one shown in Fig. 1-29 can be used. Although designed primarily for portable drill work, the 1/4-inch shank arbor adapter will fit a standard drill press geared chuck and the 5-inch rubber backing pad can be attached to the adapter by a screw and washer. Various grades of sanding paper can be installed on the pad. *Note*: Do not use a backing pad that is larger than five inches on a standard drill press.

Buffing wheels and *polishing bonnets*, 5 inches and less, can also be used on the drill press. Other valuable portable drill accessories include *rotary rasps* (Fig. 1-30), *rotary files* (Fig. 1-31), and a *flexible shaft with a chuck* (Fig. 1-32). The rotary rasps are handy for fast wood removal, slotting, and shaping, while the rotary files can be used for filing metals, elongating holes and slots, removing burrs and scale, light milling and other similar metal finishing operations. The flexible shaft permits you to employ the drill press as a power source to drill, sand and shape anywhere within the reach of the shaft. Most flexible shafts are about 40 inches long.

Vises and Hold-Downs. To drill a small piece with a drill press, either hold the

Fig. 1-30: Various types of rasps suitable for drill press work: (Top) rotary rasps; (center) disk rasp and drum rasp; and (bottom) speed rasp.

17

Fig. 1-31: Rotary files suitable for metal work.

work in a drill-press vise of some type or clamp the work securely to the table. C-clamps (Fig. 1-33A) are excellent for holding small, flat workpieces and, in many cases, for securing a long, unwieldy piece to the table to assist the left hand in holding it. For another useful clamp see Fig. 1-33B.

There are also several special clamps on the market, such as the one shown in Fig. 1-33C. This hold-down clamp is excellent for sheet metal holding. The *hold-down fence* illustrated in Fig. 1-33D has the added feature of an adjustable "shoe-shaped" hold-down clamp that rides lightly on the top surface of the workpiece and prevents it from lifting off the work table when the drill is withdrawn. The fence is useful both for guiding a workpiece and positioning it. The hold-down locks the work in the desired position. It adjusts on the post to accommodate various thicknesses of work.

The three work-holding vises shown in Fig. 1-34A, B, and C, bolt or clamp to the drill-press table to securely hold small, especially metal, workpieces. The *rotary indexing table* is the best of the three. This precision machinist's tool is designed for layout, indexing, and holding for drilling, milling, etc. The table rotates 360 de-

A

B

C

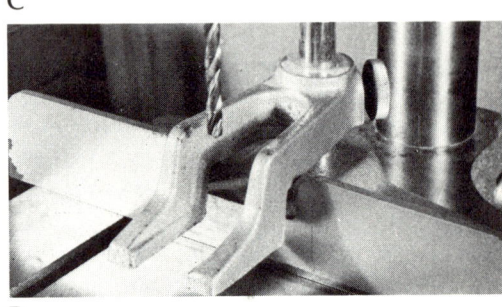

D

Fig. 1-33: Clamps and hold-downs are used to hold workpieces in place on the drill press table.

Fig. 1-32: Typical 40-inch flexible shaft with chuck.

grees. The dial is calibrated. The vise can be returned to its original setting to duplicate operations.

Lamp Attachment. Many workers want a lamp that is conveniently attached to the machine, since it is important to keep sufficient light on the work at all times. The lamp attachment shown in Fig. 1-35 is available complete with shade, socket and cord, flat links, bolts, and attachment bracket.

Multi-Speed Attachment. The multi-speed attachment mounted on a high speed model 15-inch drill press, as shown in Fig. 1-36, provides speeds ranging from 125 to 5800 rpm's with a 1725 rpm motor. With an 1140 rpm motor, on slow speed model 15-inch drill presses, there are eight speeds ranging from 85 to 3800 rpm. The attachment consists of the ball-bearing center pulley with mounting bracket, the forward V-belt from the center or jackshaft pulley to the spindle pulley and the rear V-belt from the motor pulley to the center jackshaft pulley. The regular spindle pulley and motor pulley that come with the drill press are used as a part of the drive mechanism with the attachment (Fig. 1-37).

Another multi-speed attachment available for some 15-inch slow-speed models

Fig. 1-34: Three work-holding vises..

Fig. 1-35: Parts of a typical lamp attachment and how it is installed.

features a twelve-speed range from 185 to 4600 rpm. It operates in the same manner as the eight-speed attachment.

Tilting Table. Most home shop drill presses manufactured today do not feature a tilting work table. However, manufacturers generally offer a tilting table for their larger drill presses. The 11 by 14 inch tilting table shown in Fig. 1-38 fits any 15-inch drill press with a 2 3/4-inch column. A tilt angle scale indicates up to 90 degrees both right and left.

Usually, such tables can be tilted by loosening a nut under the table. A pin fitting through corresponding holes provides a positive stop at both horizontal and vertical positions.

An auxiliary tilt table can be easily made to fit almost any drill press. The one shown in Fig. 1-39 consists of two pieces of hardwood 3/4 x 9 1/2 x 11 inches, two swivel supports made of 1/4 inch thick plywood and two 1 1/4 x 1 3/8 inch tight pin hinges. Construction details appear in Fig. 1-40. When the jig is in use it can be fastened to the drill press table with C-clamps or

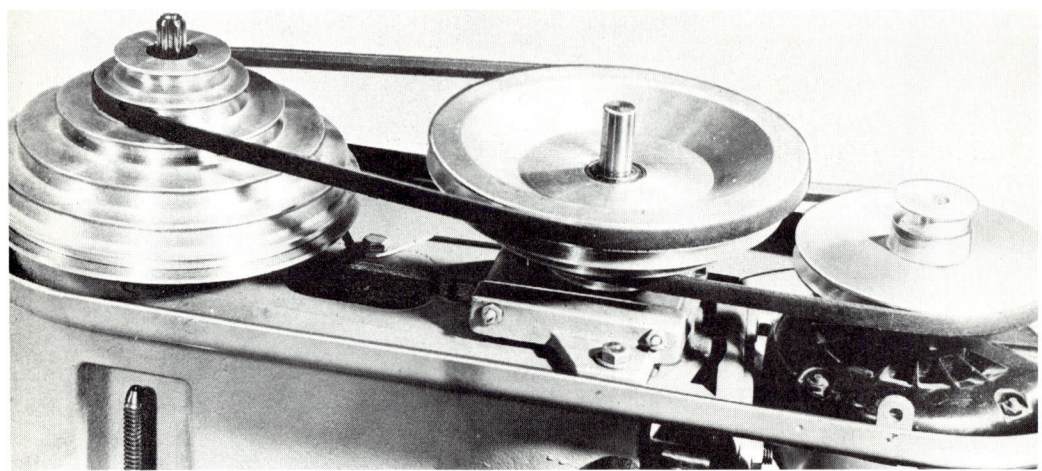

Fig. 1-36: A typical multi-speed attachment.

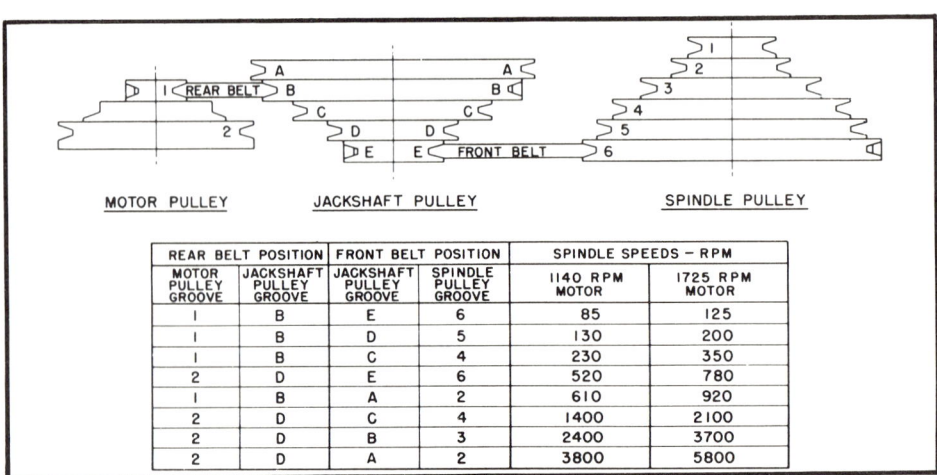

REAR BELT POSITION		FRONT BELT POSITION		SPINDLE SPEEDS – RPM	
MOTOR PULLEY GROOVE	JACKSHAFT PULLEY GROOVE	JACKSHAFT PULLEY GROOVE	SPINDLE PULLEY GROOVE	1140 RPM MOTOR	1725 RPM MOTOR
1	B	E	6	85	125
1	B	D	5	130	200
1	B	C	4	230	350
2	D	E	6	520	780
1	B	A	2	610	920
2	D	C	4	1400	2100
2	D	B	3	2400	3700
2	D	A	2	3800	5800

Fig. 1-37: How the multi-speed attachment works. As shown in the illustration, the spindle speeds would be either 520 or 780 rpms depending on the motor.

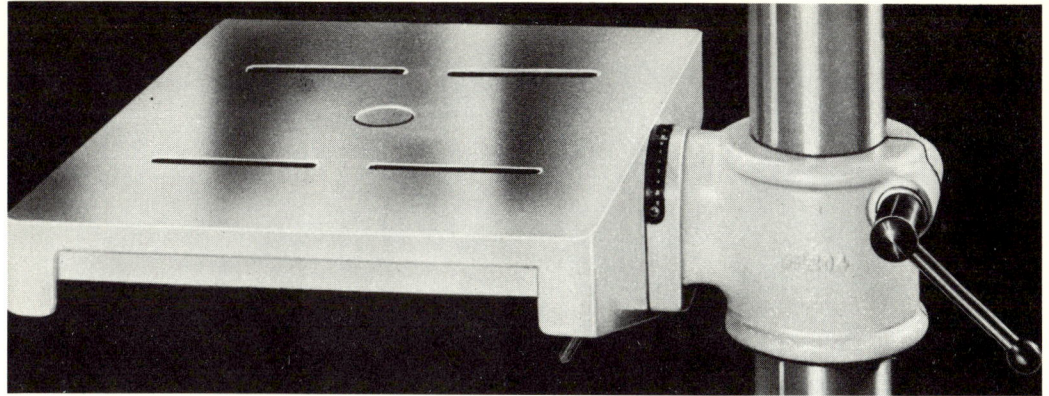

Fig. 1-38: A typical tilting table.

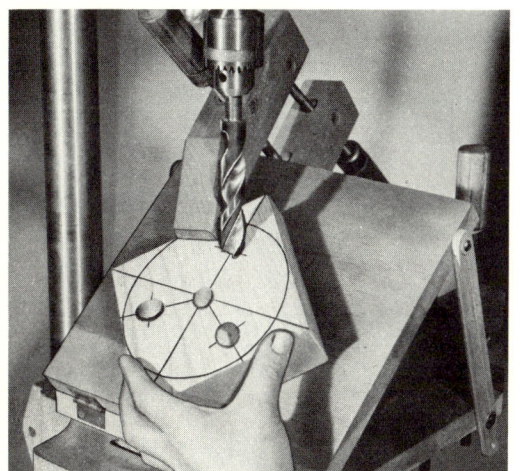

Fig. 1-39: An auxiliary drill press tilting table in use.

with four 1/4 x 2 1/4 inch carriage bolts. As with any jig, be sure to seal it with shellac or finish it with varnish or similar materials.

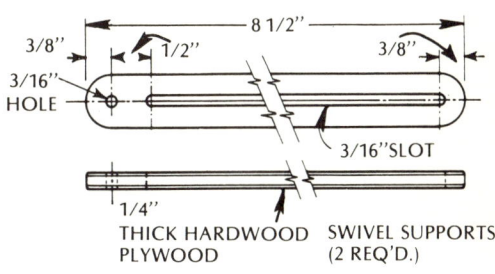

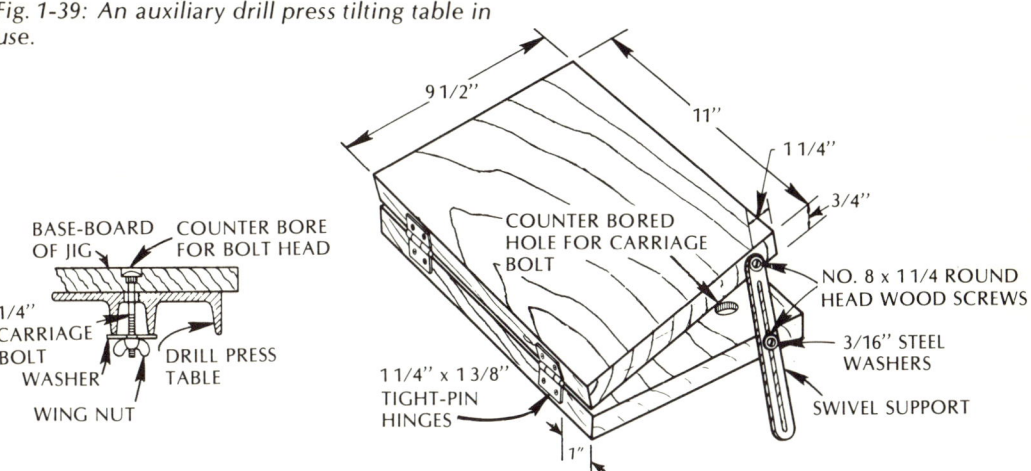

Fig. 1-40: Construction detail of an auxiliary drill press tilting table. You may wish to reduce or increase the overall width to your drill press table.

21

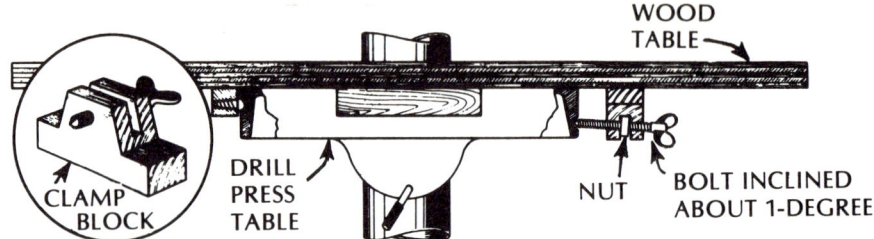

Fig. 1-41: A quick-change auxiliary table.

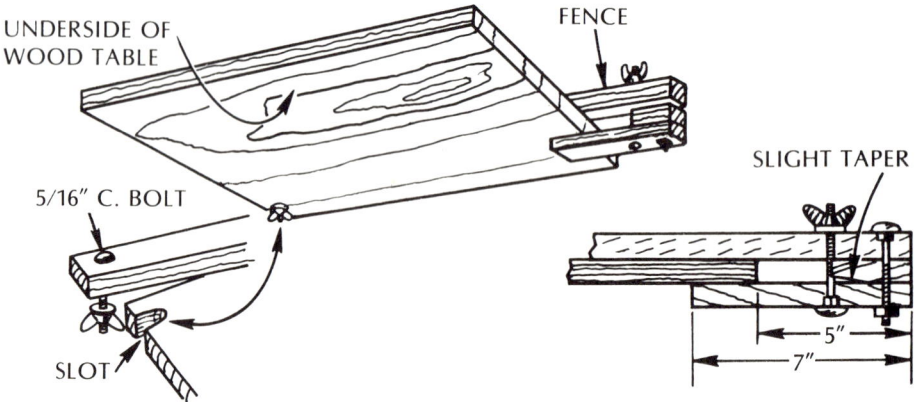

Fig. 1-42: Another quick-change auxiliary table with an adjustable table.

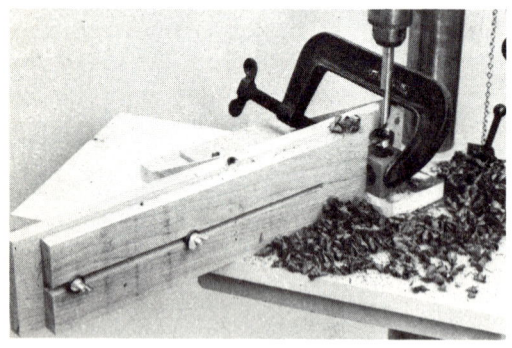

Fig. 1-43: A work-saving auxiliary table in use.

Auxiliary Wood Tables. A large working surface (generally about 18 by 24 inches) is needed for many drill press operations. You can buy a ready-made plywood or particle-board (chip-board) table or you can make it yourself. The table is bolted to the standard metal table. Because it is needed, off and on, for various jobs, a quick-change mounting is preferable. There are several ways of doing this, one method being shown in Figs. 1-41 and 1-42. This requires recessing a hole on both sides of the metal table. Another simpler way is to drill two holes in the auxiliary wood table, spacing these the same as the two slots in the table. Counterbore the two holes so that the head of a carriage bolt in each hole will be flush with the table's surface. Place two bolts in the holes and through the two slots, and fasten with washers and wing nuts.

A convenient fence to go with the wood table is detailed in Fig. 1-42. This pivots on a slot at the left end of the table, allowing the

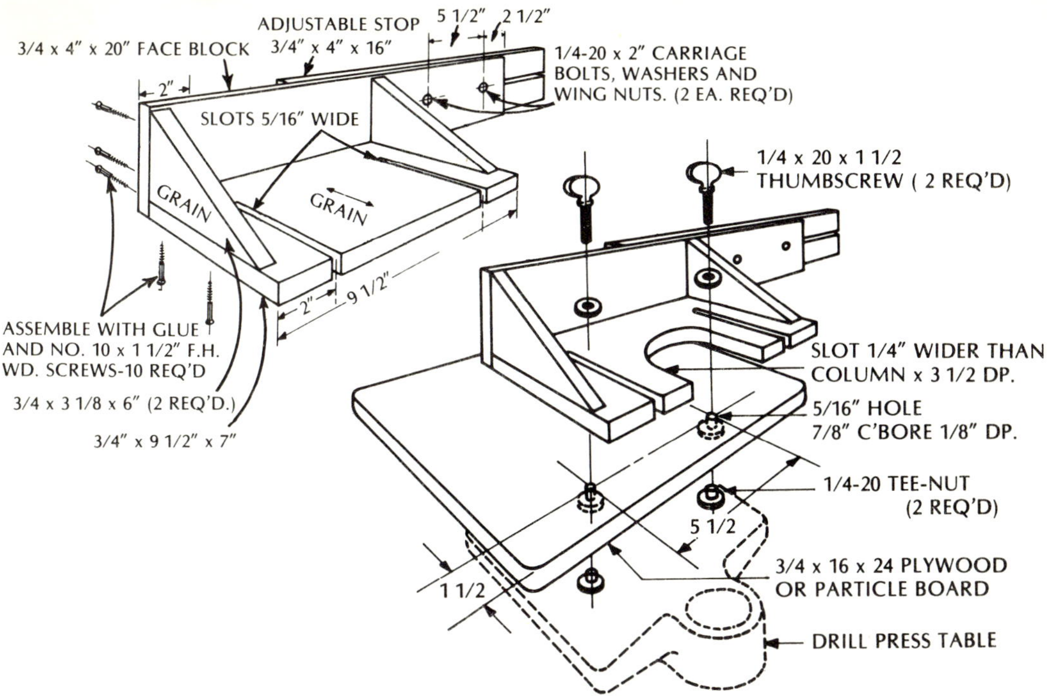

Fig. 1-44: Construction details of the work-saving auxiliary table.

fence to be moved as desired. The clamping action is obtained by turning one wing nut. When not needed, the fence can be pushed to the left, out of the slot and off the table.

Another simple auxiliary table jig shown in Fig. 1-43 saves a great deal of time by replacing blocks and clamps normally used for drill press setups. But, as with all jigs and fixtures, remember that accuracy is the key. Use good, flat boards and be sure the face blocks and adjustable stop are square with the drill press table. This table is constructed as shown in detail in Fig. 1-44.

Storage Racks. A wooden rack fitted directly to the drill press with a metal bracket will always keep your wood bits within easy reach (Fig. 1-45). As illustrated in Fig. 1-46, the 9 1/2-inch long block is drilled with 9/16-inch holes to take the half-inch shanks of the various tools. An alternate arrangement is to drill each mounting hole the same size as the tool it holds.

Fig. 1-45: Drill rack made from block of wood keeps drills ready for instant use.

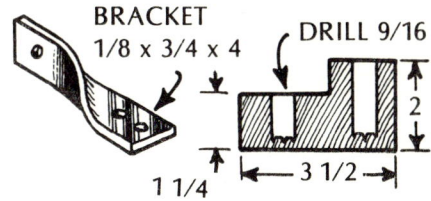

Fig. 1-46: Bracket and block for drill rack.

Another drill rack is shown in Fig. 1-47. The two tiers of this drill rack revolve independently of each other, and for easy selection of a bit or drill the whole rack will swing around the drill press column. It will stay in any position and will not interfere with the work. All three parts of the rack are made of 3/4-inch plywood cut to the dimensions shown in Fig. 1-48.

With either rack, it is a good idea to occasionally squirt machine oil in each drill or bit hole to guard against rusting.

Shop Vacuum. While not an attachment to your drill press, a shop vacuum is an accessory that will help keep your machine free of wood chips and other debris. This versatile device can also be used for other clean-up jobs around the shop and home.

Safety Goggles. Safety goggles or a safety face shield should be worn when operating any power tool and this includes the drill press.

Fig. 1-47: A column drill storage rack.

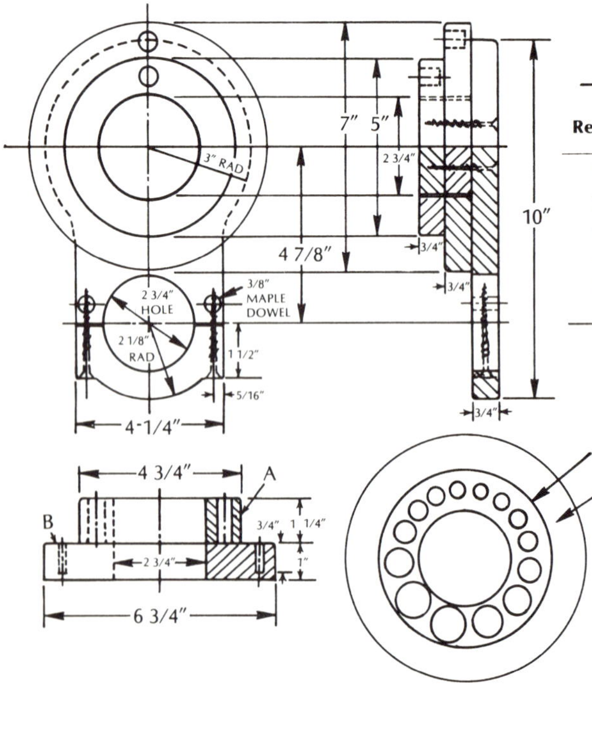

BILL OF MATERIALS

Req. No.	Name	Size
1	Base	3/4 x 6 x 10
2	Base Plugs	3/4 x 2 3/4 x 2 3/4
1	Lower Revolving Rack	3/4 x 7 x 7
1	Upper Revolving Rack	3/4 x 5 x 5
2	Maple Dowels	3/8 Diam. x 3/4
2	Flat Head Wood Screws	No. 8 x 2 1/2
2	Flat Head Wood Screws	No. 8 x 2

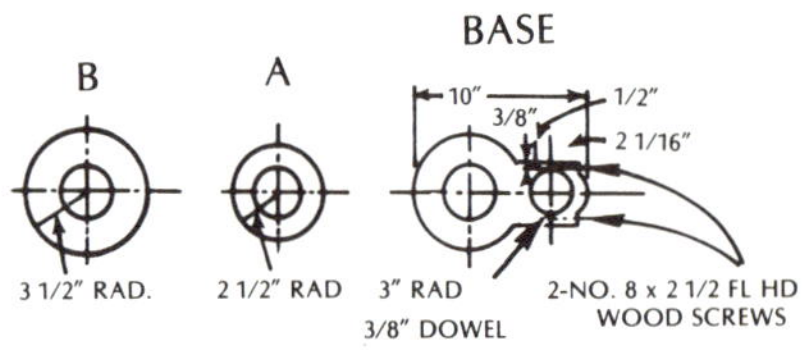

Fig. 1-48: Construction details of the column drill storage rack.

Chapter 2

BEFORE OPERATING THE DRILL PRESS

It is important to know your drill press before operating it. The information in this chapter is of a general nature, appropriate for most drill presses, including the radial type. For specific data on your drill press, carefully check the owner's manual that came with it. By using the owner's manual along with this book, you will be able to "get the most out of your drill press."

INSTALLATION

When setting up the drill press in your shop, make sure to locate it where you can easily handle any desired workpiece. Good overhead light is of utmost importance, even if the tool is equipped with a lamp attachment. As mentioned in Chapter 1, a nearby storage area to keep your bits and other accessories handy is a good step-saver to consider.

While the bench-type drill press can be mounted on the workbench, its operation will prove more satisfactory when mounted on a wood or steel stand of its own. (A steel stand can usually be purchased with the drillpress.) When mounting the drill press to the bench, use bolts and nuts to fasten the base of the machine to the benchtop. Most drill presses are mounted with three bolts. If during operation there is any tendency for the drill press to tip over, slide, or walk on the supporting surface, the stand (floor model) or bench must be secured to the floor.

On a wooden floor, the bench and stand models can be secured with lag bolts. On concrete, use masonry bolts and anchors.

The drill press is unpackaged and assembled according to the manufacturer's instructions furnished with it.

Connecting the Drill Press to a Power Source. A separate electrical circuit should be used for your power tools. This circuit should not be less than #12 wire and should be protected with a 20-Amp time-lag fuse. If an extension cord is used, use only three-wire extension cords which have three-prong grounding plugs and three-pole receptacles that will accept the tool's plug. The size of the wire will depend on the motor amperage. Replace or repair damaged or worn cord immediately. Before connecting the motor to the power line, make sure the switch is in the "OFF" position and be sure that the electric current has the same characteristics as those stamped on the motor nameplate. All line connections should make good contact. Running on low voltage will damage the motor.

The drill press must be grounded to protect the operator from electric shock. The motors for small drill presses are usually wired for 120 volts, single-phase, and are equipped with an approved three-conductor cord and three-prong grounding type plug to fit the proper grounding type receptacle, as shown in Fig. 2-1. The green conductor in the cord is the grounding wire. Never connect the green wire to a live terminal.

An adapter, shown in Fig. 2-2, is available for connecting three-prong grounding type plugs to two-prong receptacles. *This adapter is not applicable in Canada.* The green-colored rigid ear or wire, extending from the adapter, is the grounding means and must be connected to a permanent

25

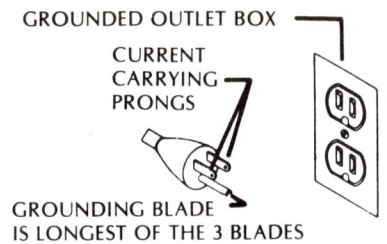

Fig. 2-1: Three-prong grounding type plug designed to fit into a grounding type receptacle.

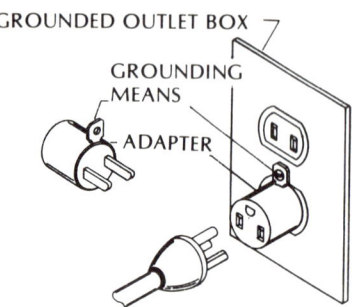

Fig. 2-2: An adapter such as shown above is suitable for connecting a three-prong grounding type plug to a two-prong receptacle.

ground, such as a properly grounded outlet box. If you are not sure that the receptacle in question is properly grounded, have it checked by a certified electrician.

DRILL PRESS ADJUSTMENTS

Most drill presses are thoroughly tested, inspected, and accurately aligned before leaving the factory. However, moving parts will wear, and the abrasive action of dust and dirt adds to this wear. Rough handling during transportation can also throw the machine out of alignment. Eventually adjustment and realignment are necessary in any machine to maintain accuracy—regardless of the care with which the tool is manufactured.

Adjusting or Changing the Spindle. While methods of interchanging spindles may vary slightly from machine to machine, the following is one of the popular arrangements. After the drill press has been disconnected from its power source, the quill is lowered to expose the collar which locks the upper end of the spindle. The collar is usually directly behind the cover plate in the front of the drill press head. This cover must be removed before making any adjustment (Fig. 2-3). After the quill is clamped in this position with the quill lock handle, the spindle is inserted at the bottom of the quill and pushed upwards, through the collar and through the drive pulley. The pulley should be slightly turned while doing this, as shown in Fig. 2-4, so that the keys in the drive pul-

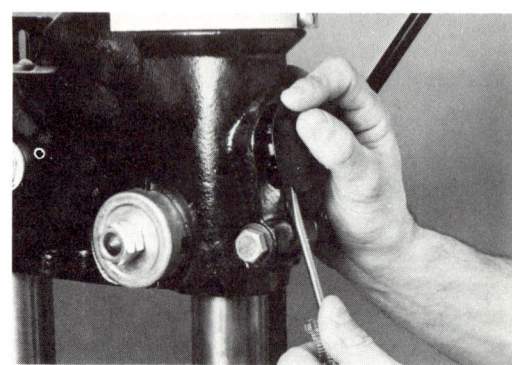

Fig. 2-3: The cover plate must be removed to get at the spindle collar.

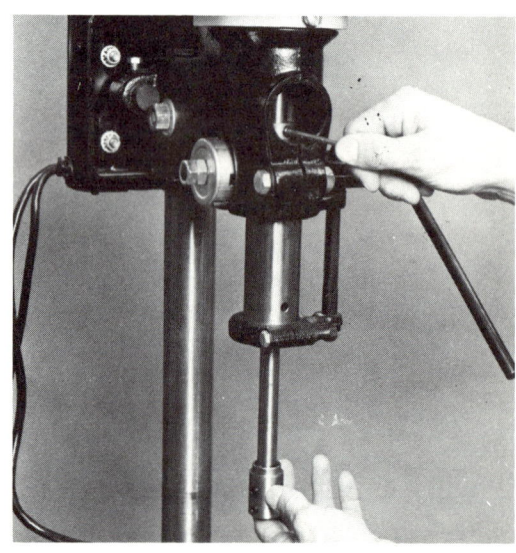

Fig. 2-4: Installing a 1/2-inch hole spindle in an 11-inch drill press.

Fig. 2-5: Various spindle adapters available for use with a 15-inch drill press.

ley will enter the keyways in the spindle. The spindle is now fully seated. The upper collar must now be locked. This is done by pushing the spindle upward and the collar downwards, so that the spindle will not slip while tightening the setscrew in the collar. The setscrew should not be tightened excessively, nor should it enter the keyway of the spindle. Either of these actions will mar the spindle and make withdrawal difficult.

Rather than replacing the entire spindle, many drill presses have various spindle adapters that can be used to hold various special bits and accessories (Fig. 2-5). Changing these adapters is simple if the manufacturer's directions are carefully followed. When attaching the adapters to the spindle, it is very important to wipe clean both the spindle taper and the tapered hole in the adapter. Then place the adapter on the spindle and tighten the locking collar. If in checking the spindle for accuracy, there should be run out, it is suggested that the adapter be removed and turned perhaps one quarter or one-half turn and replaced. This may reduce or eliminate the run-out, but it may also increase it, in which case, remove the adapter and rotate it further.

When removing the spindle in a typical radial drill press (Fig. 2-6), proceed as follows:

1. Remove the hole cover plate.
2. Run the quill down until the locking collar above the quill is accessible through the opening in the head where the hole cover plate was removed.

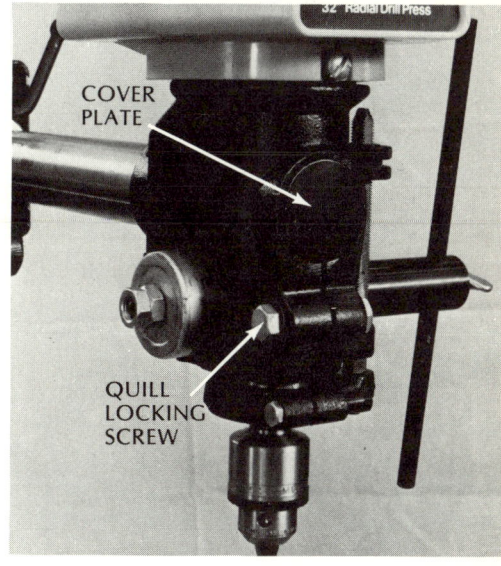

Fig. 2-6: The quill arrangement of a typical radial drill press.

3. Clamp the quill in this position and while holding the spindle to avoid dropping it, release the setscrew in the locking collar.

4. Swing the table to one side and raise the quill to the top of its stroke. This should allow sufficient clearance for lowering the spindle out of the bearings.

End play in the spindle can usually be eliminated by clamping the table at a suitable height. Lower the spindle against the table with moderate pressure while, at the same time, clamping the quill with the quill locking screw. Remove the cover plate, move the locking collar down against the top of the quill, and tighten the setscrew used to secure the collar in position. This method will remove slack in the thrust collar, thrust bearing, and quill.

Replacement of the spindle pulley bearing will usually not be necessary until the machine has had a long period of heavy use. To check the bearing, remove the spindle so that the pulley will run idle. If it turns smoothly and runs with no noise beyond the humming of the belt, the bearing is still in good condition. If it is necessary to replace the bearing, follow the instructions found in the owner's manual. Because instructions vary from machine to machine, always check the owner's manual before making any spindle adjustments or repairs.

Quill Adjustments. The quill travels in a bored hole in the head casting. These parts will remain accurate indefinitely if kept clean and lubricated according to instructions.

The spindle is raised or lowered by the hand lever. The quill can be locked at any desired point in its travel by tightening the quill locking screw or handle (Fig. 2-7). This is an especially desirable feature in operations such as sanding. Any play which might develop between the head and quill after considerable use can be taken up by partially tightening the quill locking screw.

On operations such as sanding, where

Fig. 2-7: The quill arrangement of a typical drill press. Check the owner's manual for exact locations.

the quill is clamped in place, always keep the quill as high as the work will permit so that any side thrust will be transmitted directly to the head casting.

Belt Tension. The belt should be just tight enough to prevent slipping. Excessive tension will reduce the life of the belt, pulleys, and bearings. Correct tension is obtained when the belt can be flexed about 1 inch out of line midway between the pulleys, using finger pressure. To adjust belt tension, follow the manufacturer's instructions found in the owner's manual. Generally, this is accomplished by moving the motor bracket to the required position (Fig. 2-8).

If the drill press is equipped with a 1725-rpm motor, the typical spindle speeds are approximately 600 to 5000 rpm, while the typical speeds for a 1140 rpm motor are from about 450 to 2000 rpm. The highest speed is obtained when the belt is on the largest step of the motor pulley and the smallest step of the spindle pulley (Fig. 2-9). When you change the belt position to change speed, *always disconnect the machine from its power source.*

Inserting Drills. The standard geared or

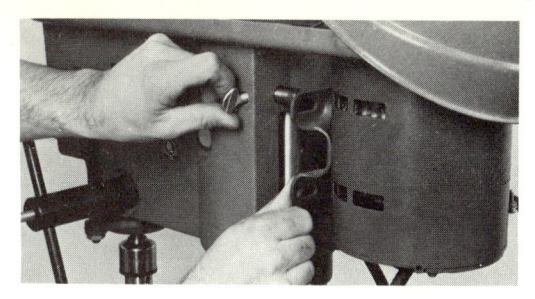

Fig. 2-8: The belt is tensioned by moving the motor bracket to the required position.

Fig. 2-9: Changing the belt position to change the spindle speed.

Fig. 2-10: Chucking a drill in a geared chuck.

Fig. 2-11: Centering the work table.

keyed chuck is opened and closed with the special wrench provided (Fig. 2-10). It can also be worked by hand although the final grip tension must be applied with the wrench. Be sure the drill's shank is centered between the chuck's jaws and it is properly secured in the chuck before the power is turned ON. Do not apply further pressure with pliers or wrenches after you hand tighten the chuck with the chuck key.

Always remove the key immediately after you use it. Otherwise the key will fly loose when the drill motor is started and may cause serious injury. Self-ejecting chuck keys are available (see page 43).

The insertion of drills and other tools in a spindle which employs an adapter to receive the shank of the drill is quite simple. The drill is pressed into the hole and the setscrews are tightened to hold it. Where taper shank drills are used, the drill is fitted by pressing it into the tapered hole at the end of the spindle, engaging the tang of the drill in the corresponding slot in the spindle. During the course of the work, the drill becomes wedged tightly in the tapered hole, and must be removed by means of a drift key.

Centering the Table. In average drilling operations, the hole in the center of the table should be directly under the drill so that the drill, after going through the work, will enter the hole in the table. Where through drilling is being done, the quill should always be brought down first without the work in place to see that the drill enters the table opening (Fig. 2-11).

Adjusting the Spindle Return Spring. For the purpose of automatically returning the spindle upward after a hole has been drilled, a coil spring enclosed in a metal case is fitted to the side of most drill presses. (Check your owner's manual for the exact details). Generally this spring is adjusted at the factory and usually requires no further adjustment. If, however, the spindle fails to return to a normal position, or if the return is too rapid, the tension should be adjusted. This is done by loosening the locknuts which hold the case in place. They should not be completely removed, but simply backed off about 1/4 inch, enough so that the case can be pulled out to clear the bosses on the head. As the case is pulled out, it must be held tightly to prevent the spring from unwinding (Fig. 2-12). The case is turned clockwise to loosen the spring; counter-clockwise to tighten it. When the quill is up, two full turns from a non-tension position should give the proper tension. Before tensioning the spring, it is well to slack it off entirely. It should be noted that the exact method of adjusting the spindle return spring varies with different models and makes of drill presses. Consult the owner's manual supplied with your machine for details.

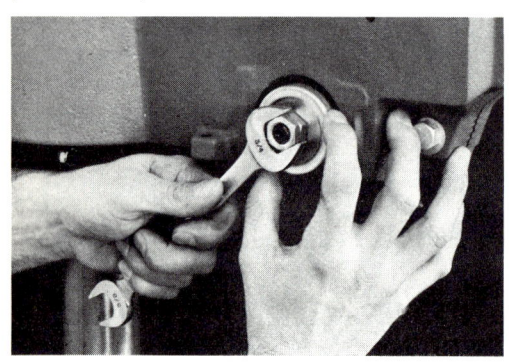

Fig. 2-12: Adjusting the spindle return spring.

Adjustment of the Radial Drill Press Head. The head of a radial drill press can be tilted right and left simply by loosening the horizontal column clamp (Fig. 2-13), and pulling out the plunger, located on the left side of the tee bracket. Tilt the head to the desired angle and lock the horizontal column clamp. A scribed line along the right side of the horizontal column is used with the degree scale on the front of the tee bracket to locate the desired setting. When returning the head perpendicular to the table, make sure the end of the plunger is engaged in the milled slot in the horizontal column, and lock the horizontal column clamp.

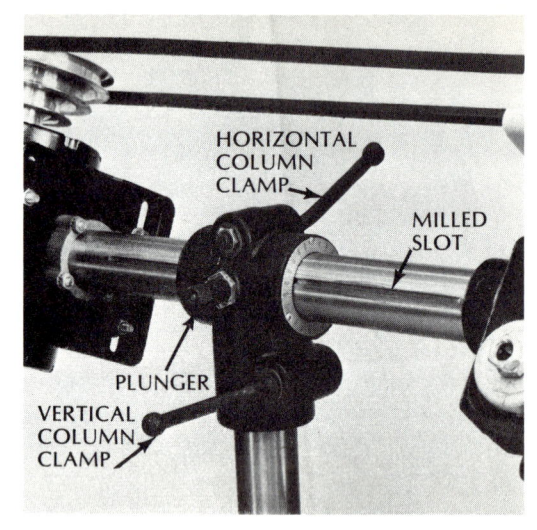

Fig. 2-13: Tilting the head of a radial drill press.

The head can be swiveled 360 degrees around the column simply by loosening the vertical column clamp, rotating the head and tee bracket to the desired position, and tightening the vertical column clamp.

Check to see if the head is set perpendicular to the table by placing a drill bit in the chuck and a combination square on the table, as shown in Fig. 2-14. This adjustment should be performed with the plunger locator engaged in the milled slot

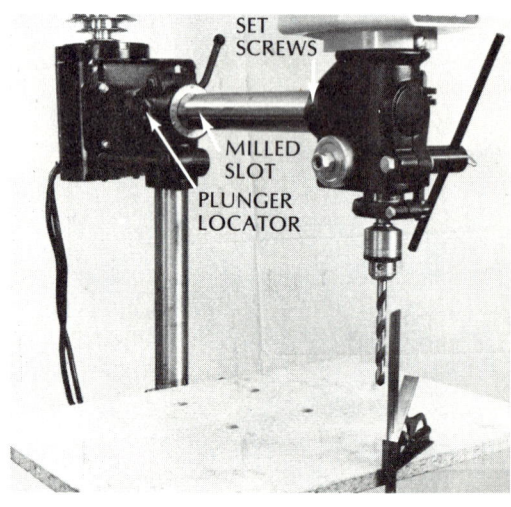

Fig. 2-14: Adjusting the head of a radial drill press perpendicular to the table.

in the horizontal column. If an adjustment is necessary, loosen the two setscrews that hold the head to the horizontal column, tilt the head until the drill bit is perpendicular to the table, and tighten the two screws. Remember that correct belt tension should be maintained while this adjustment is being made.

Lubrication. Some drill presses are not fitted with a sealed spindle bearing. Such a non-sealed bearing will need a few drops of light machine oil once or twice a month (Fig. 2-15). The oil holes in the front of the quill near each end are usually accessible when it is run down, clamped, and the front cover plate of the head casting removed. (Check your owner's manual for exact details.) The lower hole lubricates the thrust bearing in addition to the lower spindle bearings. Lubricate the surface of the quill occasionally by applying a drop or two of oil inside the head casting. Also oil the quill rack, the return spring, the pinion shaft (where it enters the head casting), the upper end of the spindle above the pulley, the flat surfaces of the stop rod, and the interior of the chuck. Be careful to keep the drive belt free of oil.

Fig. 2-15: Typical spindle lubrication hole. It is pointed out by an arrow.

With the radial type of drill press, apply a coating of white petroleum jelly to the horizontal column occasionally to allow it to move freely.

On all drill presses, a coat of paste wax or a rub-down with a piece of wax paper will protect the surface of the table. Wiping with a slightly oiled cloth will discourage rusting of the column and quill.

LAYING OUT THE WORK

Practically every hole that is drilled, requires a layout mark which will locate either its approximate or exact position. The easiest and most accurate method of marking a hole location is to draw intersecting lines that show you the center of the hole. To accomplish this, various tools are used in making the layout, ranging from a pencil, scriber, square, hammer, and punch to quite expensive instruments essential for very exacting work. Let us take a look at some of the more common layout tools.

Rules. Of all measuring tools, the simplest and most common is the steel (or fiberglass) rule. This rule is usually 6 or 12 inches in length, although other lengths are available. Steel or bench rules, as they are sometimes called, may be flexible; but the thinner the rule, the easier it is to measure accurately, because the division marks are closer to the work.

Generally, a rule has four sets of graduations, one on each edge of each side. The longest lines are the "inch marks," representing inches. On one edge of one side, each inch is divided into 8 equal spaces, so each space represents 1/8 inch. The other edge of this side is divided into sixteenths. The 1/4- and 1/2-inch marks are commonly made longer than the smaller division marks to facilitate counting, but the graduations are not numbered individually. They are sufficiently far apart to be counted without difficulty. The opposite side is similarly divided into 32 and 64 spaces per inch, and it is common practice to number every fourth division for easier reading. An important point about rules is illustrated in Fig. 2-16: Dimensions should be taken off with the rule on edge to avoid possible errors as shown in the lower portion of the drawing.

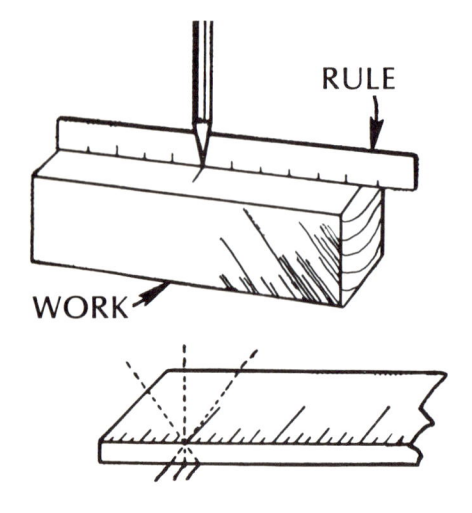

Fig. 2-16: How dimensions should be taken off with a rule.

There are many variations of the common rule. Sometimes the graduations are on one side only. Sometimes a set of graduations is added across one end for measuring in narrow spaces, and sometimes only the first inch is divided into sixty-fourths, with the remaining inches divided into thirty-seconds and sixteenths. Steel machinist's rules, which are popular variations of the common rule, are thin blades of steel of varying lengths, widths, and thicknesses, usually graduated in inches and various subdivisions of the inch on each edge of both sides (and often at the ends). The makers label the various subdivisions of the inch with graduation numbers. For example, the No. 4 graduation has sixty-fourths on the first edge, thirty-seconds on the second edge, sixteenths on the third edge, and eighths on the fourth edge. By means of sliding or fixed attachments, a great variety of length measurements can be made with the ordinary steel rule. The U.S. Customary System of Units (U.S.C.S.) rules offer a great variety of both fractional and decimal increments as fine as one one-hundredth of an inch (0.010 inch), which is about the limit of the resolving power of the naked eye. Metric scales are normally graduated by millimeters (0.039 inch) and half millimeters (approximately 0.020 inch). Most manufacturers now make many of their linear measuring devices with both metric and U.S.C.S. graduations. All dimensions in this book follow the U.S.C.S. However, a table of inch/millimeter conversion is included on page 119.

Awls and Scribers. While pencil, crayon, and chalk are common marking "tools," the most popular tools for shop use are the awl and scriber. When using a pencil for marking, it should have medium-hard lead and always be sharp.

The *awl*, or *scratch awl* as it is often called, is a sharp-pointed piece of steel, like an ice pick, which is used to score wood and metal (Fig. 2-17A). It is also used for making pilot holes when boring in wood.

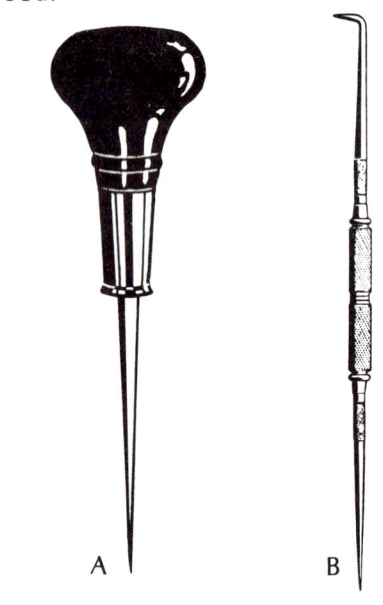

Fig. 2-17: Two common marking tools.

The scriber (Fig. 2-17B) is used for marking lines on metal, especially when you are measuring with a rule. Keep the scriber sharp, and use it like a pencil, with only enough pressure to make a clear mark.

Various substances must be used to mark on other surfaces. A pencil makes a very clear mark on wood but is almost invisible on polished steel. If the work is rough iron, a coating of white or blue stick chalk will usually do the trick. After the chalk is rubbed with the fingers for better distribution, the coating will show pencil or scribed marks clearly. A similar coating is made from powder chalk or whiting mixed with water to form a solution which is brushed on the work (Fig. 2-18). A strong solution of copper sulphate is excellent for polished steel. This is applied with a swab and builds up a nice copper surface in a few minutes. A 20-percent solution of silver nitrate is preferred by some workers for copper and brass. It is also possible to purchase various kinds of ready-mixed coating solutions. For most occasional work, an ordinary pencil or a scriber will make a mark strong enough to allow the work to be done. The intersection of the scribed lines is usually indented with a punch, and the punch mark rather than the scribed mark is the guide for drilling the hole.

Combination Squares. One of the most useful tools for average work is the combination square. The square head may be adjusted to any position along the 12-inch blade and clamped securely in place. Since the blade is graduated in inches and fractions of inches, the combination square can also serve as a depth gauge, height gauge, or marking gauge. Two of the faces of the head are ground at right angles to each other, and a third face is ground at 45 degrees. This means that the combination square can be used as an inside and outside try square, as well as a miter square (Fig. 2-19). A small level is built into the head for checking whether surfaces are level or plumb, and a small scribe is housed in a hole in the end of the head for marking layout lines. Without the square head, the blade can be used as a straightedge.

A typical application of a combination square is centerlining. When one or more holes are to be located between two parallel surfaces (Fig. 2-20), the combination square is used to mark two lines on the workpiece equidistant from the two parallel sides. If accurate measurements are made these two lines should coincide; if not, they should be close enough so that it will not be difficult to "spot" the center between them. In fact, even if the sides are not parallel, when the two lines are drawn as close together as possible, it should permit the spotting of a center between the lines.

Most combination squares have a center head that can be slid onto the blade in place of the square head. This is a V-shaped member designed so that the center of the 90-degree V lies exactly along one edge of

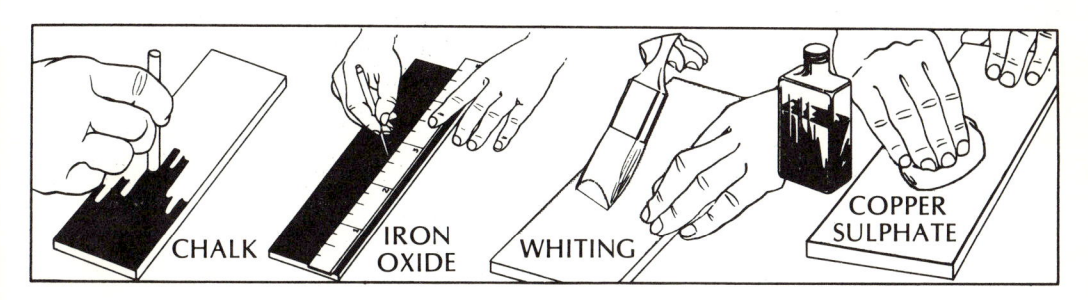

Fig. 2-18: Coatings for various markings.

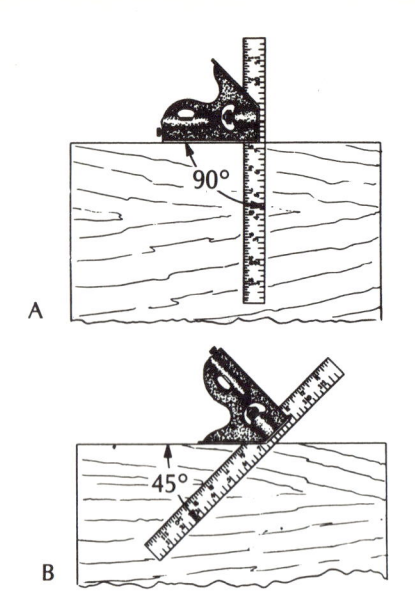

Fig. 2-19: (A) Squaring a line on the stock; (B) Laying out a 45-degree angle.

the blade. This attachment is useful when locating the exact center of round stock. That is, any two lines scribed along the inner edge of the blade will intersect at the exact center of a truly round piece (Fig. 2-21).

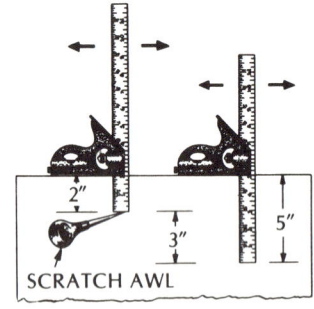

Fig. 2-20: Laying out holes between two parallel surfaces with a combination square.

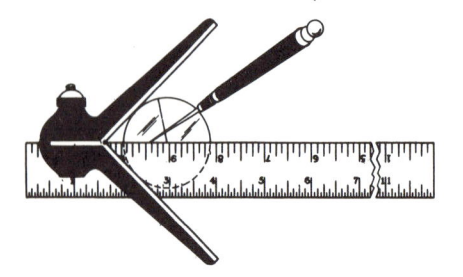

Fig. 2-21: Using a combination square to find the center of round stock.

The exact center of a square or rectangular workpiece can be found by marking diagonals between opposite corners (Fig. 2-22). These lines will cross at the center of the square or rectangular workpieces. If the piece, however, is neither square nor rectangular, the center can be found by drawing lines that will cross near the center (as closely as possible), using the 90-degree square or center head of the combination square, then spotting the center.

Protractors. The protractor is useful for establishing angles and angular lines. Some combination squares are equipped with a protractor head which can be used as shown in Fig. 2-23 to layout angular lines.

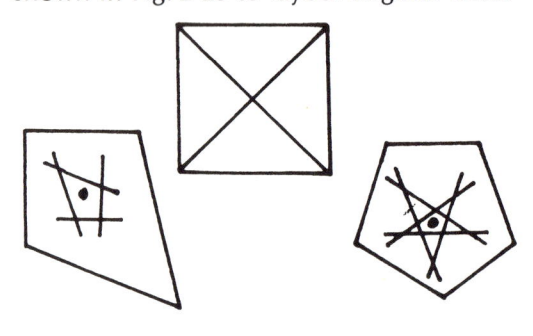

Fig. 2-22: Centering square or rectangular workpiece ends as well as odd-shaped pieces.

Another type of protractor is illustrated in Fig. 2-24. When using this type of protractor to space three holes equidistantly around a center, for example, draw one radius (from the center) and measure from the center on this radius to locate the center of the first hole. Then, with the protractor set at 120 degrees (1/3 of 360 degrees) and using the same radius, locate the two other holes.

Dividers. Of the many uses for dividers, the most common are accurately stepping off measurements several times and scribing circles.

To lay out a circle with dividers, set the dividers at the desired radius, using a rule as shown in Fig. 2-25A. Note that the radius being set here is being taken at a central

34

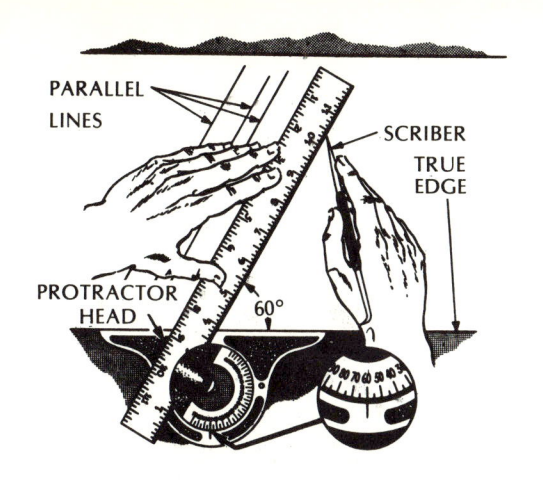

Fig. 2-23: Drawing angular lines with a protractor head.

portion of the rule, rather than at the end. This reduces the chance of error, as each point of the dividers can be set on a graduation. Place one leg of the dividers at the center of the proposed circle, lean the tool in the direction it will be rotated, and rotate it by rolling the knurled handle between your thumb and index finger (Fig. 2-25B).

Dividers can also be used to establish hole centers with equal-distance spacing along a given centerline (Fig. 2-26).

Calipers. The most common calipers used for drill press work are hermaphrodite types. This type of calipers has one bent leg which rides against the edge of

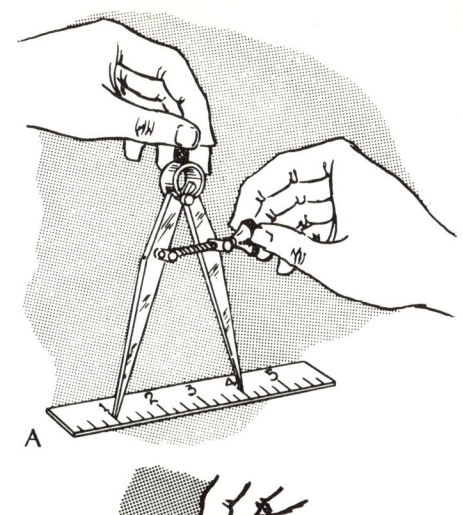

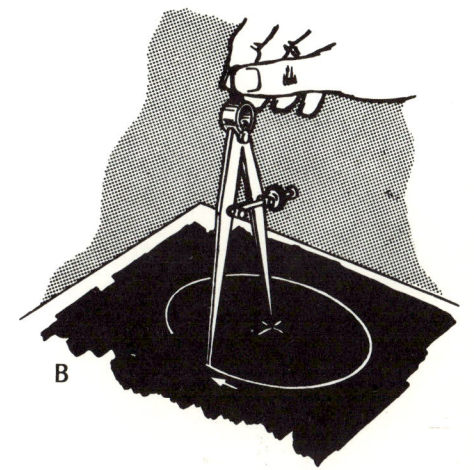

Fig. 2-25: Laying out a circle with dividers.

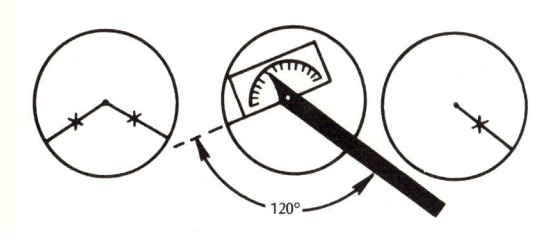

Fig. 2-24: Plotting holes with a simple protractor.

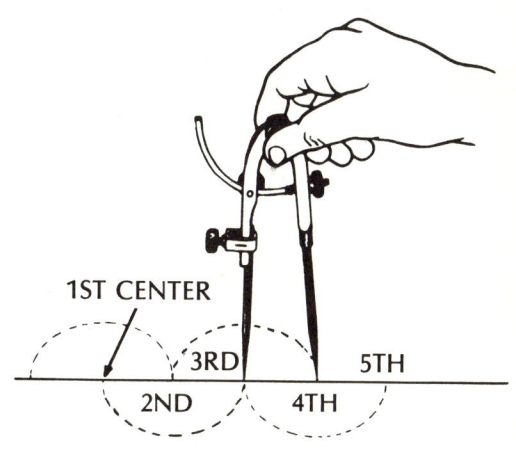

Fig. 2-26: Transferring hole center with dividers.

the work, while the other leg, fitted with a scriber, makes the layout marks. A typical layout operation can be seen in Fig. 2-27.

Gauges. Gauges are measuring devices. They are special tools used because they are more convenient to handle or easier to read than the conventional rules. Of course, there are many types of gauges. Some are standard and are used frequently metal-shop or woodworking-shop projects. Others are more specialized and are used only by the advanced artisan and the professional.

A surface gauge, for example, is used to transfer measurements to a workpiece by scribing a line, and to indicate the accuracy or parallelism of surfaces. It consists of a base and an adjustable spindle, to which a scriber or an indicator can be clamped (Fig. 2-28). Surface gauges are made in several sizes and are classified by the length of the spindle, the smallest spindle being 4 inches, the average 9 or 12 inches and the largest 18 inches. The scriber is fastened to the spindle with a clamp. The bottom of the base has a deep V-groove, to allow the gauge to be seated on a cylindrical surface.

The spindle of a surface gauge may be adjusted to any position with respect to the base and tightened in place with the spindle nut. The rocker adjusting screw provides for fine adjustment of the spindle by pivoting the spindle rocker bracket. The scriber can be positioned at any height and in any desired direction on the spindle by tightening the scriber nut. The scriber may also be mounted directly in the spindle and nut mounting, in place of the regular spindle, and used where working space is limited or the height of the work is within the range of the scriber.

To set a surface gauge for height, first wipe off the top of a layout table or surface plate and the bottom of the surface gauge. Use either a combination square or a rule with a rule holder to get the measurement.

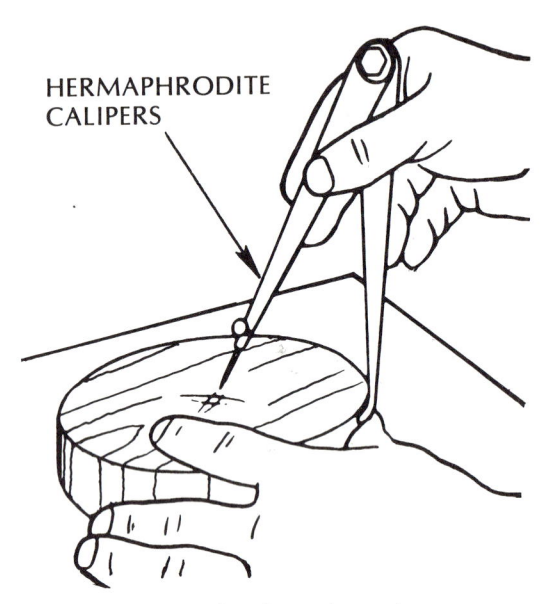

Fig. 2-27: Hermaphrodite calipers in use.

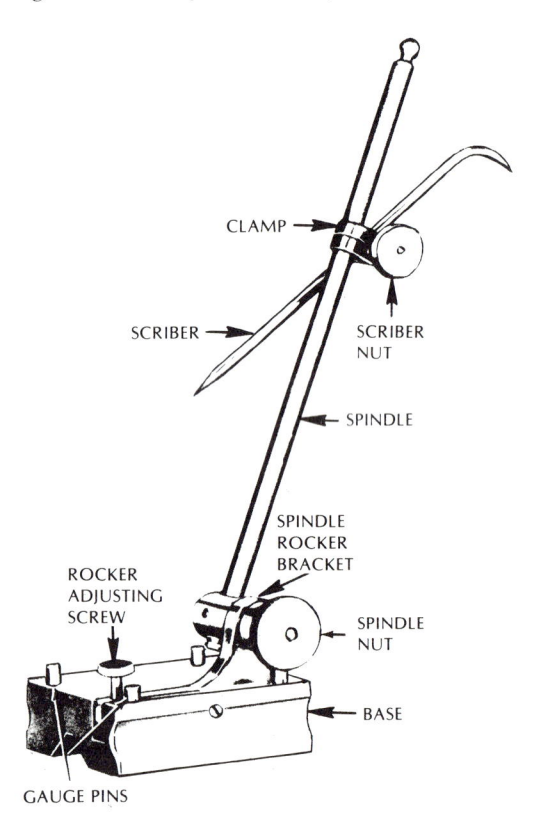

Fig. 2-28: Parts of a surface gauge.

A rule alone cannot be held securely without wobbling, and consequently an error in setting generally results. Because a combination square is generally available, its use for setting a surface gauge is explained in this section.

Place the square head of a combination square on a flat surface (Fig. 2-29) and secure the scale so that the end is in contact with the surface. Move the surface gauge into position and set the scriber to the approximate height required, using the adjusting clamp that holds the scriber onto the spindle. Make the final adjustment for the exact height required with the adjusting screw on the base of the gauge.

Punches. After the scriber or pencil has been used to locate the hole position, it is necessary to further indent this point. This is done with a center punch (Fig. 2-30). There are various sizes of center punches; the size selected will depend upon the work. Some common punch styles are shown in Fig. 2-31. At the extreme left is a medium-size center punch. Next is a fine center punch. The two examples at the right are spacing punches, which can be set so that when one leg is placed in a previously punched hole, the second leg can be punched to lay out a hole at a certain exact distance from the first hole. This type of punch is often quite useful on complex layouts where a large number of small holes must be spaced at regular intervals.

Automatic center punches are useful for layout work. They are operated by pressing down on the shank by hand. An inside spring is compressed and released automatically, striking a blow on the end of the punch. The impression is light but adequate for marking, and serves to locate the point of a standard punch when a deeper impression is required. When using any punch, the first indentation should be made lightly. Then, if accurate work is being done, take a pair of dividers and set the dividers to the radius of the

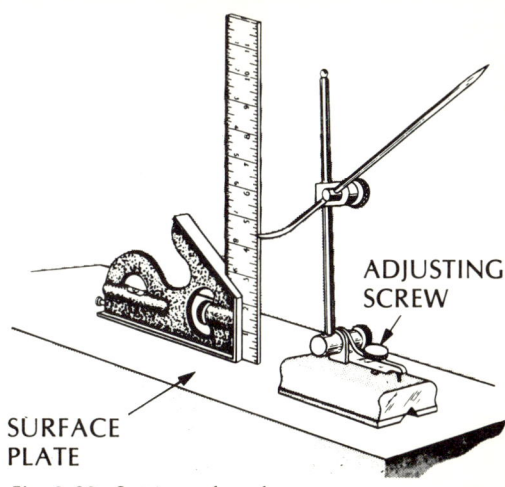

Fig. 2-29: Setting of surface gauge.

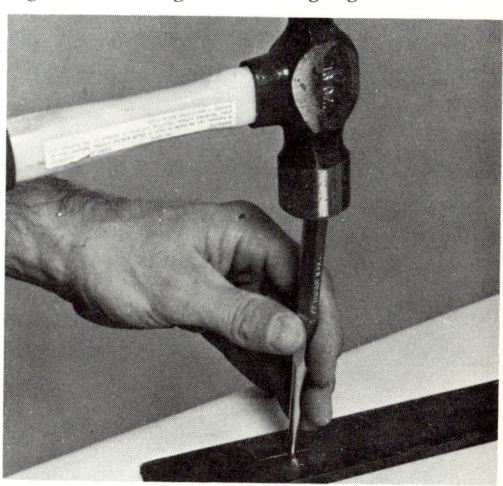

Fig. 2-30: Center punching should be done carefully so that the punch mark comes at the exact intersection of the layout lines.

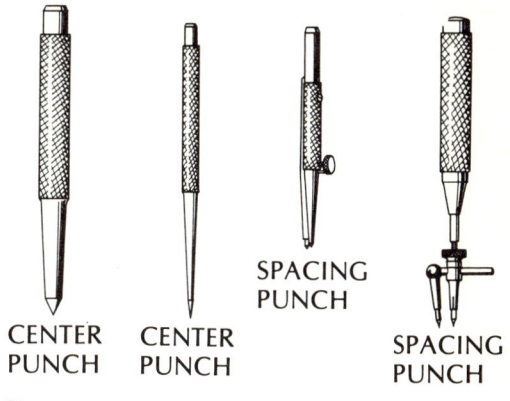

Fig. 2-31: Various styles of punches.

hole to be drilled. Scribe an outer circle around the punch mark (Fig. 2-32), and then scribe another circle about half the diameter of the first. Prick-punch lightly the intersections of the large circle with the layout lines, as shown; then deepen the center punch mark. A layout such as this is a positive check against poor work since the circles immediately show any fault, giving the operator plenty of opportunity to adjust the hole location.

There are two things to remember when striking a punch with a hammer:

1. When you hit the punch, you do not want it to slip sideways over your work.

2. You do not want the hammer to slip off the punch and strike your fingers.

You can prevent both of these possibilities by holding the punch at right angles to the work and then striking it squarely with your hammer. *Remember to wear safety goggles.*

Laying Out Dowel Holes. Dowel holes are similar to other kinds of holes as far as layout is concerned, except that the mating holes must match. Various methods for laying out dowel holes are shown in Fig. 2-33.

A simple but accurate method is shown in Fig. 2-33A. Ordinary pins are stuck in a block of wood and placed in such a position as to come between joining members, as shown. When the joining members are pushed forcibly together, the pinheads make an impression on each piece which serves as a guide when drilling the dowel holes.

Double-point thumb tacks can be used in the same manner. The more standard method of working, however, is to use dowel centers or "pops" as shown in Fig. 2-33B and C. Here, after drilling the holes for the dowels in one piece of wood, you insert dowel centers in these holes. Then you align the two pieces of wood as they will be joined. When you press them together, the points on the dowel centers mark the second piece of wood. It is now possible to drill holes at these center marks. When the pieces are connected with dowels, the blind dowel joint is perfectly aligned. Dowel centers come in assorted sizes to fit holes from 1/8 to 1 inch in diameter.

When locating dowel holes in a series of boards that will be joined edge-to-edge, position the board edges and butt them surface to surface. Using a combination square (Fig. 2-33D), mark the hole location on one edge and carry the line across all the pieces. Identify the board faces which will be at the top after assembly. Drill dowel holes in the edges at cross-lines, using an arrangement, such as the one described in Chapter 3, that will gauge holes automatically.

When doweling joints other than those at edges, it is a good idea to make a template of stiff cardboard, thin plywood, hardboard, or even sheet metal if the long-term use justifies it. Drill 1/16-inch or smaller dowel center holes at the desired locations. Locate the template accurately first on one piece, then the other, to mark dowel centers with a center punch or awl (Fig. 2-33E).

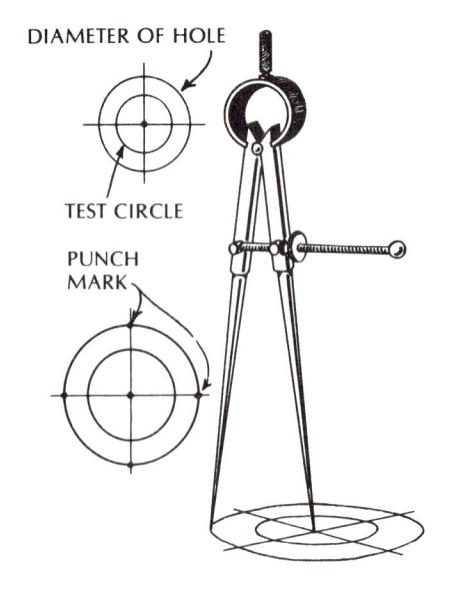

Fig. 2-32: Use of double circles for large holes.

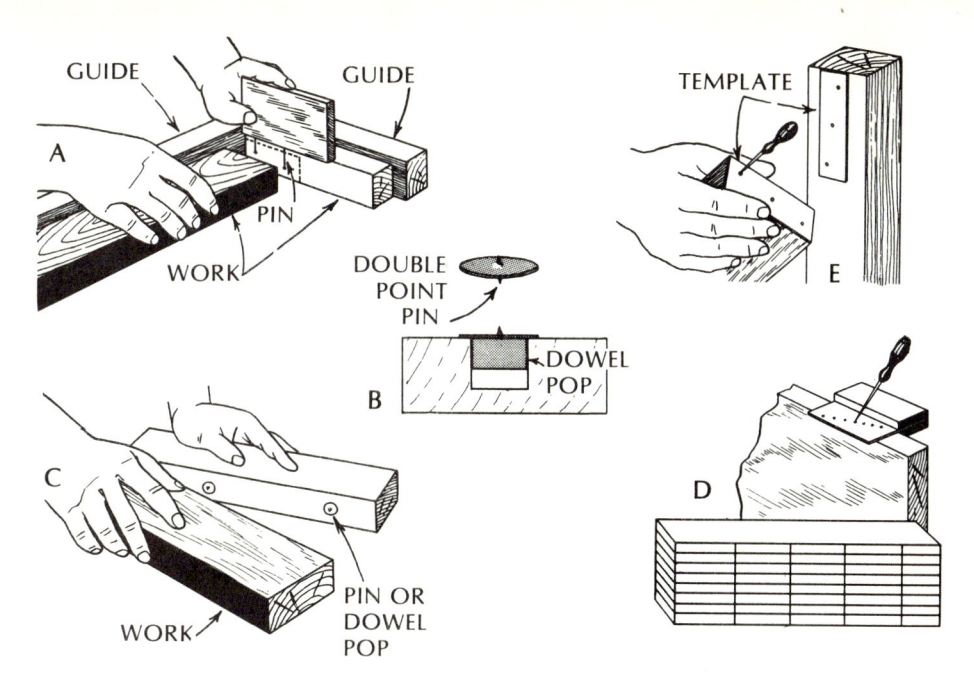

Fig. 2-33: Various methods used for locating dowel holes.

The template marking can also be used when a number of pieces are to be drilled alike. Also remember that certain pieces of hardware, such as a hinge, hasp, or drawer pull, provide their own template.

Time spent in making a good layout is usually worthwhile.

HOLDING THE WORKPIECE

You can hand-hold wood and softer materials if size permits and if the hole size is not excessive. But, as stated in Chapter 1, it is usually wise to clamp or lock the workpiece to the table. By using either clamps or a vise, you have the assurance that, should the cutter grab in the hole, particularly at the breakout point, the work will not be twisted out of grasp. Such a possibility must be considered if you do not wish to jeopardize your fingers.

There are many fixtures and jigs that make the task of holding a workpiece simple. For example, round work can be held easily with either a commercial or homemade V-block (Fig. 2-34A). Exact center holes can be located by placing the block

Fig. 2-34: Holding round work in a V-block (A) and a clamp (B).

39

so that the drill point centers in the bottom of the "V" when the quill is run down. A C-clamp can also be used to hold round work (Fig. 2-34B).

A fence on the drill press table will serve as both guide and safety device. When the fence is in place, any twisting force exerted by the drill will be taken by the fence, not your hands. You can make your own fence by clamping a straight strip of wood to the table (Fig. 2-35A), or you can use one of the accessory fences made specifically for the drill press (Fig. 2-35B). The commercial fences have the added feature of an adjustable "shoe-shaped" hold-down clamp that rides lightly on the top surface of the workpiece and prevents it from lifting off of the work table when the bit or drill is withdrawn (Fig. 2-36). Other fixtures and jigs for holding work will be described in the later chapters of this book.

SPEED AND FEED

Factors which determine the best speed to use in any drill press operation are: kind of material being worked, size of hole, type of drill or other cutter, and quality of cut desired. The smaller the drill, the greater the required rpm. The speed should be higher for soft materials than for hard ones.

On most drill presses, it is impossible to get the *exact* recommended speed, but you can come close by adjusting the drive belt on the step-cone pulleys. You will find a chart giving the various speed ratios available with your particular drill press somewhere in the owner's manual or you can use the chart given on this page as a general guide. Regardless of the speed selected, remember that the tool should cut steadily, smoothly, and without excessive vibration, no matter what the material. Decreasing or increasing rpm's is sometimes necessary because of differences in boards, even from the same type of wood. After some experience with your machine, you will know which pulley step is best in each case.

SUGGESTED SPINDLE SPEEDS—RPM

DRILLING SPEEDS

Hole Size—Inch*	Softwoods	Hardwoods	Plastics	Aluminum	Brass	Cast Iron	Mild Steel
1/16	4700	4700	4700	4700	4700	4700	2400
1/8	4700	4700	4700	4700	4700	2400	1250
3/16	4700	2400	2400	4700	2400	2400	1250
1/4	2400	2400	2400	4700	2400	1250	700
5/16	2400	1250	1250	2400	1250	1250	700
3/8	2400	1250	1250	2400	1250	700	700
7/16	2400	1250	1250	1250	1250	700	
1/2	1250	1250	1250	1250	700	700	
5/8	1250	700	700	700			
3/4	1250	700	700	700			
7/8	1250	700					
1	700	700					
1 1/4	700	700					
1 1/2	700	700					
2	700	700					

*For intermediate sizes, use speed suggested for next larger hole.
Use slower speeds for deep holes or if drill bit burns or melts material.

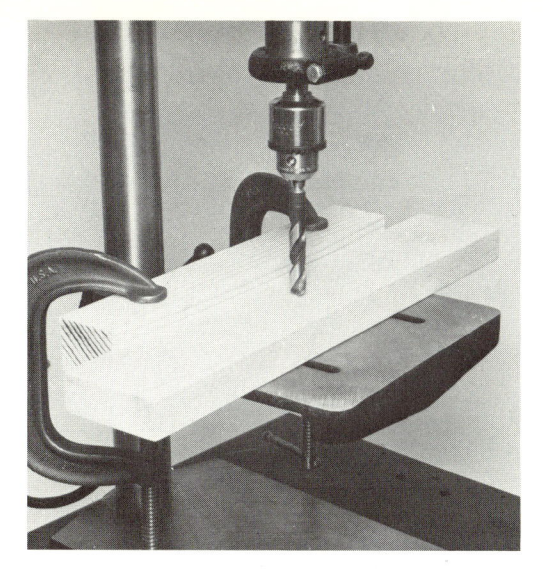

Fig. 2-35: (A) A straight strip of wood clamped to the table makes a good fence; (B) a typical commercial fence.

Feed is the amount of pressure you apply to control penetration. Too much pressure will force the tool beyond its cutting capacity and result in rough cuts and jammed or broken tools. Too light a feed, particularly with metal or other hard material, causes overheating of the tool and burning of the cutting edge. The best results will be obtained by matching the correct speed with a steady feed pressure that lets the tool cut easily at an even rate. The proper feed and speed makes the job easier.

Fig. 2-36: An adjustable "shoe-shaped" hold-down clamp is frequently used in mortising work.

TROUBLESHOOTING GUIDE FOR THE DRILL PRESS

In spite of how well a drill press is maintained, problems do come along. The following troubleshooting guide will help solve the more common problems.

Trouble: *Drill press will not start.*
Probable Cause
1. Drill press not plugged in.
2. Fuse blown or circuit breaker tripped.
3. Cord damaged.
4. Overload relay not set.

Remedy
1. Plug in drill press.
2. Replace fuse or reset circuit breaker.
3. Replace cord.
4. Push overload reset button.

Trouble: *Overload kicks out frequently.*
Probable Cause
1. Extension cord too light or too long.
2. Feeding drill too fast.
3. Cutting tools in poor condition (dull).
4. Low supply voltage.

Remedy
1. Replace with adequately sized cord.
2. Feed drill more slowly.
3. Replace or resharpen cutting tools.
4. Contact your electric company.

Trouble: *Spindle does not come up to speed.*
Probable Cause
1. Belt slipping.
2. Motor not wired for correct voltage supply.

Remedy
1. Adjust belt tension.
2. Refer to motor nameplate for proper wiring.

Trouble: *Spindle runs in wrong direction.*
Probable Cause
1. Motor runs in wrong direction.

Remedy
1. Refer to motor nameplate for proper wiring.

Trouble: *Motor overheats.*
Probable Cause
1. Extension cord too light or too long.
2. Excessive feed pressure.
3. Low supply voltage.

Remedy
1. Replace with adequately sized cord.
2. Feed cutting tool more slowly.
3. Contact your electric company.

Trouble: *Machine vibrates excessively.*
Probable Cause
1. Drill press not mounted securely on stand, floor, or work bench.
2. Faulty V-belt or damaged pulley (s).
3. Bent motor shaft, or off-center tool in spindle.

Remedy
1. Tighten all mounting hardware.
2. Replace V-belt or pulley (s).
3. Replace motor or tool.

Trouble: *Drills unsatisfactory holes.*
Probable Cause
1. Improperly sharpened tools.
2. Loose quill.
3. Tool not properly chucked.
4. Improper speed.

Remedy
1. Sharpen tools.
2. Adjust quill clamping screw.
3. Chuck tool properly.
4. Check cutting tool chart, and adjust speed.

Trouble: *Quill will not return.*
Probable Cause
1. Broken or improperly adjusted return spring.
2. Tight quill bore.

Remedy
1. Replace or adjust return spring.
2. Adjust quill and clamping screw, and lubricate quill.

Trouble: *Excessive end-play in spindle.*
Probable Cause
1. Loose locking collar at upper end of spindle.

Remedy
1. Adjust locking collar.

Trouble: *Holes not perpendicular to table (radial drill press).*
Probable Cause
1. Spindle not perpendicular to table.

Remedy
1. Adjust head.

SAFETY WITH THE DRILL PRESS

A few safety precautions must be remembered during drill press operation. They are:

1. Know your drill press. Read the owner's manual very carefully. Learn its applications and limitations, as well as the specific potential hazards peculiar to it.
2. Use safety goggles or a face shield.
3. Be sure the chuck key is removed from the chuck before turning on the power. Using a self-ejecting chuck key (Fig. 2-37) is a good way of insuring that the key is not left accidentally in the chuck. Also to avoid accidental starting, make sure the switch is in the "OFF" position before plugging in the cord. *Always disconnect the drill from the power source when making repairs.*
4. Never attempt to use a hand auger bit in a drill press. Use only drills and bits designed for machine use. Consult the owner's manual for recommended accessories. The use of improper accessories may cause hazards.
5. Hold the work firmly so that it will not fly or spin off the table. It is generally best to fasten the piece securely with clamps, or hold it in a vise. This is especially true when drilling or boring small pieces.
6. Keep the guard on the spindle pulley to prevent your hair and clothing from getting caught. In fact, no loose clothing, gloves, neckties, or jewelry should be worn when working on the drill press. A hair net is recommended for long hair.
7. Use the recommended spindle or chuck. Most operations can be done successfully with the 0 to 1/2-inch-capacity geared drill chuck.
8. Be sure the drill bit or cutting tool is locked securely in the chuck. Remember that all adjustment should be made with the power off.

Fig. 2-37: Two types of self-ejecting chuck keys.

9. Adjust the table so that the hole in the table center is beneath the drill, or set the depth stop to avoid drilling into the table. It is a good idea to place a piece of wood beneath the work to prevent this.

10. Do not use too high a spindle speed. Stay as close to the recommended speeds as possible. If there is any doubt, use the lower speed. The wrong application of high speed can burn up the cutting tool and/or workpieces, and can hurl the work off the table with considerable force. Too slow a speed with a heavy feed can cause the tool to dig into the workpiece, which can stall the motor or break the cutting edges. *Always disconnect the machine from the power source when changing speeds or making adjustments.*

11. On deep cuts, raise the bit frequently to clean the chips out of the hole. If the drill becomes stuck in the hole, turn off the machine before attempting to raise the bit.

12. Use a brush to keep the table and workpiece free of sawdust or chips. Always disconnect the machine from the power source before cleaning.

13. When using sanding drums and other abrasive accessories, make sure the area is well ventilated.

14. Never try to stop the machine by grabbing the chuck after the power is turned off. Do not run the tool unattended. Turn the power off and do not leave the drill press until the chuck comes to a complete stop.

15. To make your drill press kidproof, it is a good idea to lock the tool in the "OFF" position using a padlock (Fig. 2-38).

Fig. 2-38: One of the best ways to "kidproof" a drill press.

Chapter 3

BORING IN WOOD

Boring in wood or drilling in metal is probably the first thought that comes to mind when considering the operations that can be performed on a drill press. Boring is usually very simple, as most holes are bored at right angles to the surface. It is therefore necessary only to clamp the work to the table, select the correct bit, and set the depth. The general procedure discussed here should be followed before boring a hole.

1. Select the correct kind of bit or drill and fasten it securely in the chuck. To bore holes that are 1/2-inch or less in size, use a twist drill. Holes from 1/2 to 1 1/4 inches can be cut with a variety of boring tools including auger bits, speed bits, twist drills, and Foerstner bits. Holes larger than 1 1/4 inches are best cut with an expansive bit, hole saw, or circle cutter.

2. Adjust the drill press for the correct cutting speed. As mentioned in Chapter 2, this speed will vary with the size of the hole, depth of the hole, kind of wood, and the type of bit. As a general rule, the softer the wood and the smaller the cutting tool, the higher the speed. See the chart on page 40 as a guide to setting cutting speeds.

3. Make certain the layout has been properly made and that the center lines of the holes are well marked. Usually starting holes are not needed in wood when boring holes of less than 1/4-inch. Make sure the bit enters the wood on the marked center line. It is wise to provide a starting hole for the point of the bit when the holes are larger than a 1/4-inch, if the wood being bored is coarse grained, when accuracy is important, or if the surface to be bored is at an angle. Generally, this starting hole can be made with an awl. Make the hole just big enough for the point to enter. Do not make it too large. When a bored hole must be enlarged (Fig. 3-1A), plug the original hole (Fig. 3-1B) so that you can center the bit for the new hole.

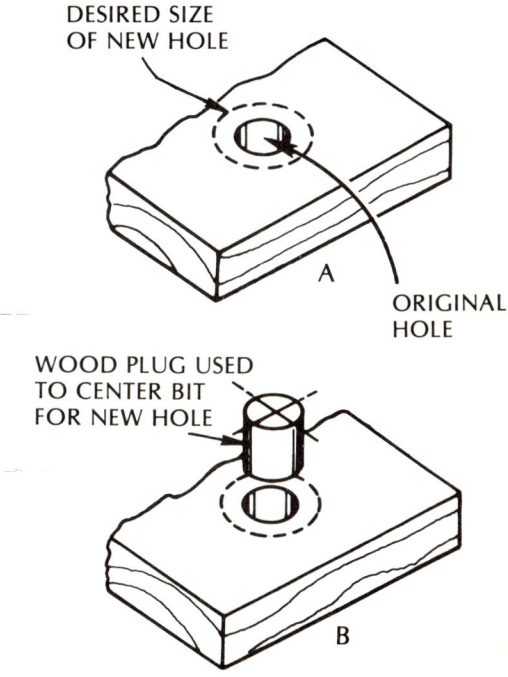

Fig. 3-1: Method of enlarging an already bored hole.

4. The bit should be centered over the center hole of the table (Fig. 3-2). If the bit is too large to enter this hole (as with a hole saw), place a scrap block (preferably 1/2-inch or thicker) under the workpiece. It is a good idea to use a base block in any case. This will prevent splintering at the underside of the work. With a base block, drilling can be done over any part of the table (Fig. 3-3).

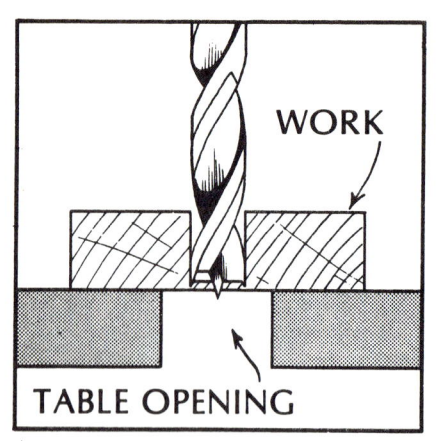

Fig. 3-2: Center the work over the table opening.

Fig. 3-3: A base block prevents splintering on the underside of the work.

5. Clamp the work securely when necessary, especially when using cutting tools with only one cutting edge, such as a fly cutter, since they have a greater tendency to rotate the work.

BORING A HOLE

When boring a through hole, as mentioned earlier, it is a good idea to use a scrap block of wood under the workpiece to prevent splintering the underside when the bit breaks through. An alternate method of preventing splintering is to bore the hole almost—but not entirely—through, and then turning the work over to finish the hole from the opposite side.

When boring, the feed pressure should be regulated by feel. Do not force a bit.

Employ only enough pressure to keep the bit cutting. When very hard wood is bored, make certain that the bit does not overheat. Back the bit out of the hole frequently during the boring, to permit it to cool. If the wood is green, its chips tend to gum up and clog a drill bit. When boring such wood, occasionally back the bit out, *stop* the drill press, and clean the chips from the bit before continuing the operation.

Holes can be bored to a specified or predetermined depth in either of two ways. Both make use of the stop-rod nuts which most drill presses have. In the first method, after one hole has been bored to the desired depth, the lower stop nut is set against the lug on the head through which the stop rod passes. Return the quill to the up position and tighten the upper stop nut against the lower stop nut and all subsequent holes will be bored to exactly the same depth (Fig. 3-4).

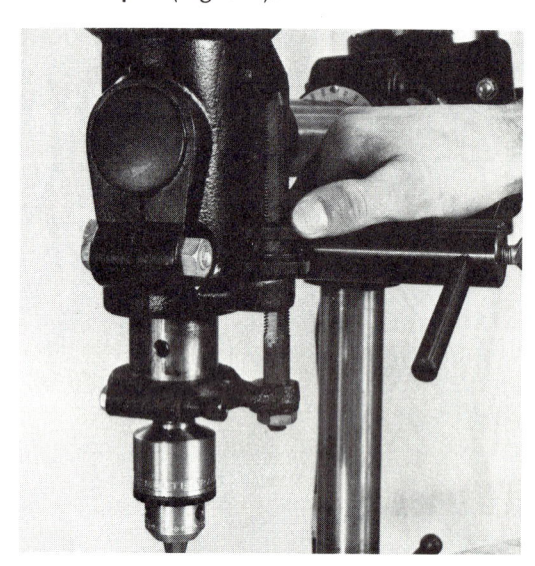

Fig. 3-4: Adjusting the stop nuts to bore holes of exactly the same depth.

Another way of boring holes to the exact depth is to mark on the side of the workpiece, indicating the hole depth (Fig. 3-5). Then extend the quill so the drill point touches the mark and set the stop-rod nuts as described previously.

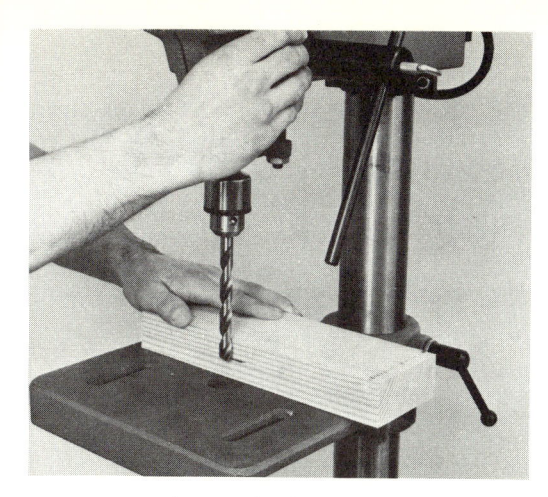

Fig. 3-5: Another method of boring holes to the same depth.

Boring Deep Holes. Various definitions are given for how deep a deep hole is. The average drill press has about a 4-inch quill travel; the most common size of machine bit also has a usable length of 4 inches. Therefore, holes deeper than 4 inches are classified as deep holes.. The simplest method of boring deep holes is to work from opposite ends of the workpiece. If the hole is less than twice the quill stroke, one of the following methods can be employed.

In the first method, the properly sized drill is chucked and used to spot a hole in an auxiliary wood table (Fig. 3-6A), or a piece of scrap wood clamped to the drill table. The drill table is now lowered the required distance to accommodate the workpiece. To align the hole in the auxiliary table or under the spindle, it may be necessary to remove the bit and replace it with a metal rod or piece of dowel stick (Fig. 3-6B). Replace the bit and bore the hole in the workpiece to the maximum depth. Then cut a short piece of dowel rod as a guide pin, and place it in the hole in the auxiliary table. Turn the work to be bored over the guide pin and then bore from the other end (Fig. 3-6C).

If your drill press has a tilting table, it can be used to advantage in end-for-end boring. In this second method, fasten an auxiliary fence or jig such as a V-block to the table and then place the table in a vertical position. The location for the hole must be aligned with the bit (Fig. 3-7) before boring. Bore from both ends.

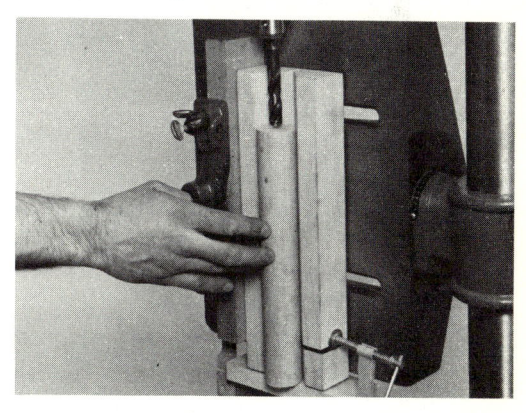

Fig. 3-7: Boring deep holes in a vertical position.

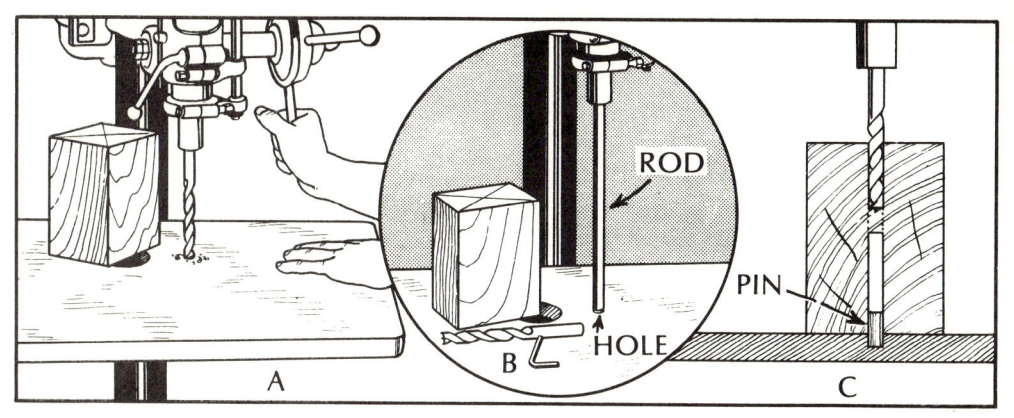

Fig. 3-6: Guide pin method of alignment for end-for-end boring.

Where a longer twist drill or a similar extra-length bit that is longer than 4 inches is being used, the bit may be long enough to go through the work, but the operator must still contend with the distance of the quill travel of the machine. One method of working is shown in Fig. 3-8. The table is vertical and the work is clamped securely in place. The end of the work is supported on a second drill table. The first hole is made to the maximum depth. The supporting table and clamp are then reset, and the second maximum bore is made. This continues until the hole is complete. A simpler method of working is shown in Fig. 3-9. Here, the first full stroke is made, by sinking the drill into the work, to the full depth of the stroke. The feed handle is then released, keeping the drill in the work. Turn off the machine. The table can then be raised to support the work in its new position, or a base block can be slipped under the work, as shown. For example, with a quill stroke of 4 inches and a bit length of 6 inches it is possible to drill the entire hole.

When end-boring a long piece, do not attempt to balance it on its end. Instead, use a simple jig such as the one shown in Fig. 3-10A. The jig consists of two or more pieces of scrap lumber, which act as a butt against the tool column. The jig is then clamped near the top of the workpiece on

Fig. 3-8: One method of deep hole boring.

Fig. 3-9: With a drill longer than 4 inches, the work is bored to the maximum feed depth and then advanced by blocking or raising the table.

A

B

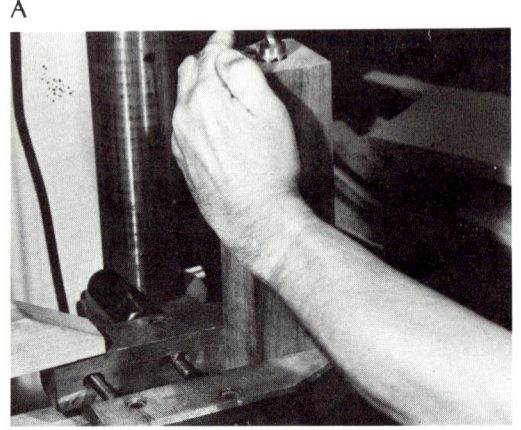

Fig. 3-10: Methods of holding long pieces for end boring.

the table (or tool base). Hold the work and jig squarely against the tool column and begin boring. If more support is necessary, use a second jig near the bottom of the workpiece. Handscrew clamps (Fig. 3-10B) can also be used to hold long work.

In all deep hole drilling, cutting should not continue after the flutes of the bit have passed below the work surface. When the flutes are below the work surface, the chips cannot get out, and burning starts immediately. Clearing the chips by lifting the drill frequently is always good practice even when the flutes are above the work surface.

The best drills for deep hole work are those having a uniform diameter from end to end. Given an accurate start, such a drill will bore a straight hole. Double flute and solid center bits are often poor performers since they are commonly back-tapered from the cutting lips, and this may cause the bit to drift. Spur bits and hollow augers do the best work. Extension bits are not generally recommended for home shop work since they have a tendency to whip.

Boring Large Holes. Holes of more than a 1 1/2-inch diameter can be classified as large holes. The removal of a comparatively large amount of wood when boring these holes puts a considerable twisting strain on the work. For this reason, it is advisable to clamp and support the work when using any kind of bit of more than 1 1/2 inches (Fig. 3-11A). This particularly applies to any kind of bit which has only one cutting edge. The four most popular large hole cutters (see page 6 for more details) are as follows:

1. Multi-spur bits are considered moderate-size cutters which bore plywood without tearing. Figure 3-11B shows a multi-spur bit in action. They will also bore at an angle. This syle of bit is also useful for cutting segments from the sides of the work (Fig. 3-12), a job that could only be done otherwise by clamping two pieces together, and drilling down into the joint.

2. The expansive bit (with the screw threads removed from the point) is another moderate-size cutter that is good if the hole is to be blind.

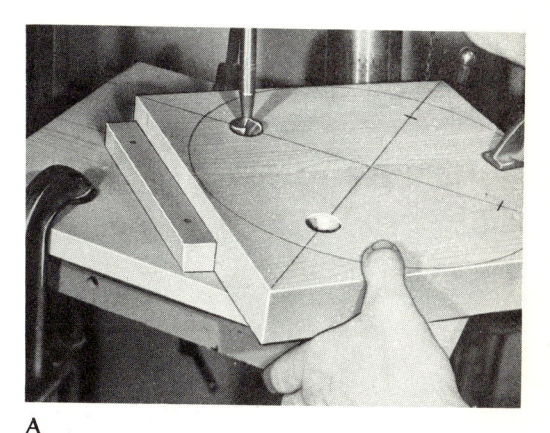

Fig. 3-11: When boring large holes be sure the work is well clamped (A). A multi-spur bit in action (B).

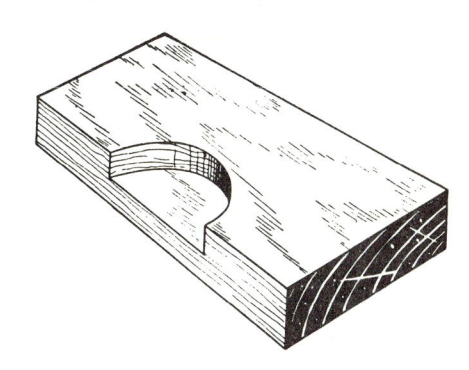

Fig. 3-12: How a segment can be cut with a multi-spur bit.

3. Hole saws (Fig. 3-13) or fly cutters will remove plugs having a 1/4-inch hole in the center, but they cannot bore a blind hole. When using a hole saw, the mandrel drill point must extend beyond the tip of the hole saw teeth by 1/16 of an inch or more. Start the hole saw square to the workpiece surface with steady feed pressure. Unbalanced tooth engagement results in erratic hole saw action.

4. Circle cutters (Fig. 3-14) bore like hole saws, but at a greater diameter. When employing a circle cutter, keep in mind that the cutter end should be ground to an 80-degree angle to prevent tearing the wood.

Large hole boring, regardless of the type of cutter employed, requires special precautions. In addition to clamping the work well, use lower speeds than the normal ones specified on page 40. In fact, when using the circle cutter (regardless of the size for which it is set), employ only the lowest speed. Also remember that the outboard cutting end of a flycutter can be just a blur *even* at slow speeds, therefore your hands should be kept away from the work.

Boring Holes in Series. The easiest method of spacing holes equally is to accurately measure the distance between the holes with a rule and mark off the space with a pencil (Fig. 3-15). A wood fence clamp fastened to the work table can be used to establish the desired edge distance (Fig. 3-16).

The use of a stop pin simplifies the job where a number of equally spaced holes must be bored. As shown in Fig. 3-17 the pin works through a block, and is set in the previously drilled hole to locate the position of the next hole. The distance between the first two holes must be accurately marked. A simpler setup with a nail

Fig. 3-13: A hole saw in action.

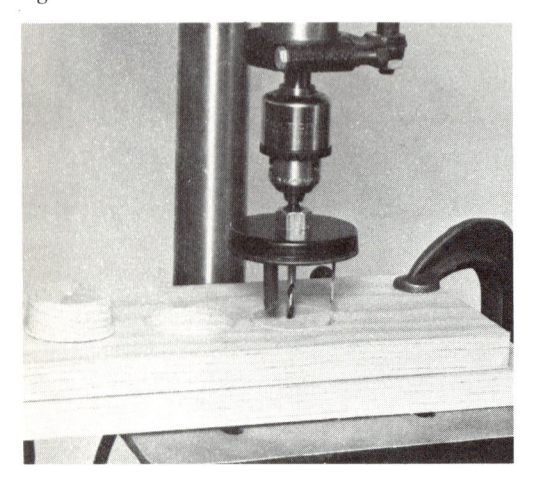

Fig. 3-14: The circle cutter in action.

Fig. 3-15: Boring a series of holes on marked work.

for the spacing pin (Fig. 3-18A) will then set the spacing to bore as many succeeding holes as desired. The nail and block idea may also be used for duplicate single holes (Fig. 3-18B).

Another way of using a guide fence when boring equally spaced holes is shown in Fig. 3-19. The fence which is clamped to the table has a series of equally spaced holes pre-drilled to the desired spacing. The fence is then adjusted with the work

A

B

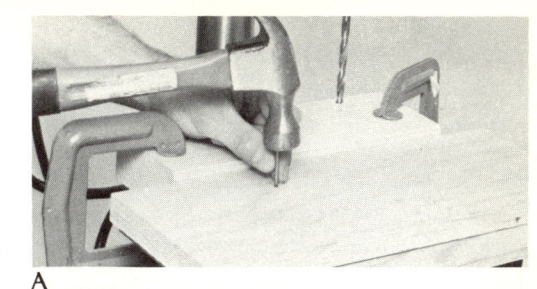

Fig. 3-18: Using a nail or nailed strips as guides.

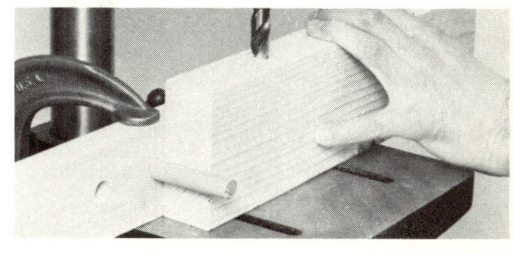

Fig. 3-19: Boring equally spaced holes using a dowel as a spacer.

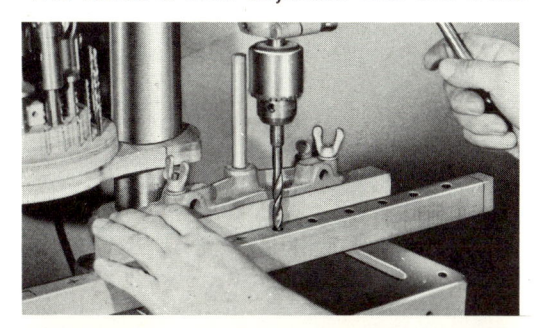

Fig. 3-16: A fence can be used to establish edge distance (top). A stop can be used when multi-drilling is done (bottom).

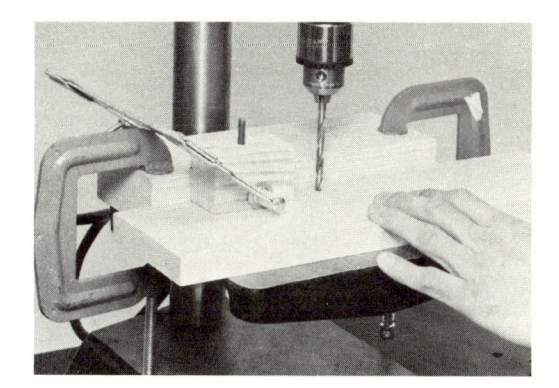

Fig. 3-17: Guide pin method of boring equally spaced holes in a series.

held against a dowel rod fastened in one of the holes. "Matching" holes will be bored correctly. The dowel rod is moved to the other holes in the fence to complete the boring operation.

The use of spacer blocks (one for each hole) is still another way of boring holes with the same spacing. The block widths should all be the same where equal spacing is desired, with different widths for uneven spacing. As shown in Fig. 3-20, the spacer blocks are positioned between a stop block and the workpiece end. After the first hole is bored, remove the spacing blocks one at a time to reposition the work for successive holes. The workpiece must be tight against the fence throughout the entire procedure.

When needed, jigs can be designed to guarantee the accurate placement of the holes. In Fig. 3-21, the hardboard boring jig and the workpiece are moved from one position to the next to accomplish the boring.

To space holes evenly in a circle requires a piece of wood to be clamped to the table (Fig. 3-22). In this case, two holes are pencil marked and bored in the same manner as is shown in Fig. 3-18, leaving the bit in the second hole. A nail is then driven through the center of work to provide a pivot point,

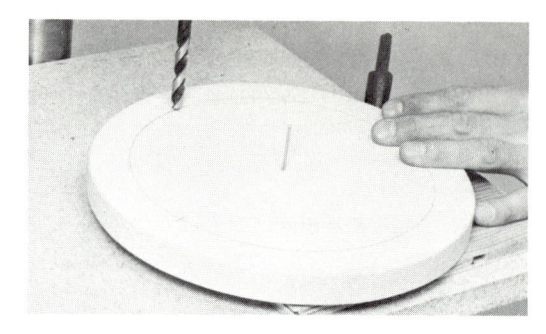

Fig. 3-22: Boring equally spaced holes in a circle.

while another nail is driven through the first bored hole. The remaining holes are then located by using the spacing pin. In circular work of this kind, the spacing of the first two holes must be exact to have the work come out even. Somewhat similar work is done with the pivot pin alone, which locates the holes accurately in relation to the center, but does not space them equally around the circle. Where a through pivot hole is not permissible, it is practical to use a short pivot pin extending about halfway into the underside of the work.

Holes in Round Work. Examples of boring in round work are shown in Fig. 3-23. All make use of the V principle of centering a round workpiece. If you have a tilting table, it can be tilted to 45 degrees (Fig. 3-23C) so that its surface and the surface of the fence form the V-shape necessary. V-blocks should be adjusted carefully. The bit must enter the highest point of the work to be accurately in line with the center. If the bit is brought down and the point of the V-block centered under the drill point, any round work placed in the V will automatically center (Figs. 3-23A and B).

Flat work can also be mounted by gripping the work in a drill vise or a wood handscrew. Either of these devices will hold the work upright, eliminating the need to tilt the table. For occasional work, it is possible to simply hold the work erect while the hole is being drilled.

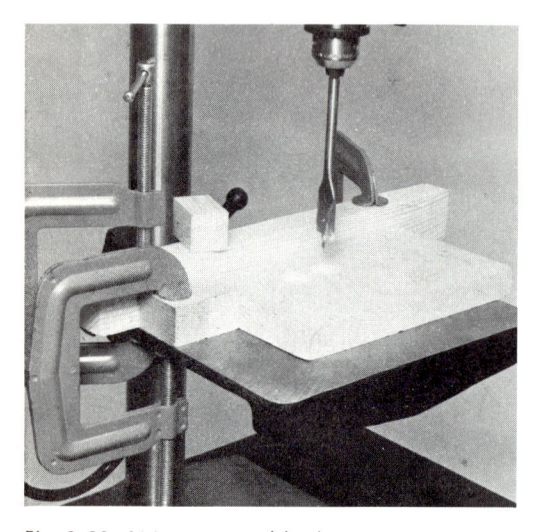

Fig. 3-20: Using spacer blocks.

Fig. 3-21: Using a boring jig.

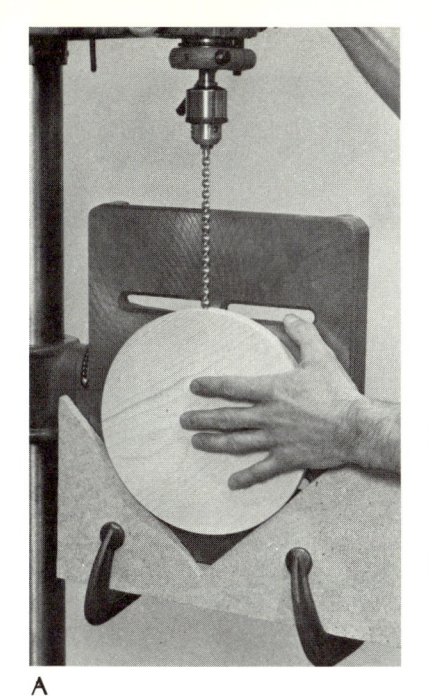

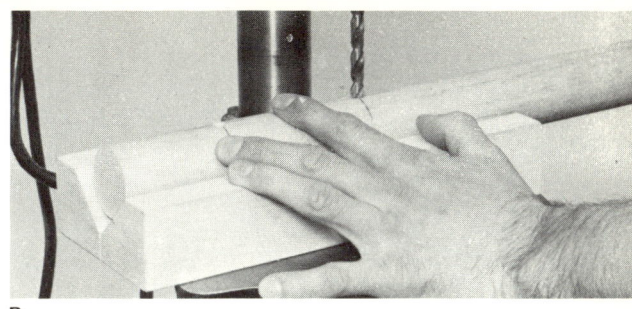

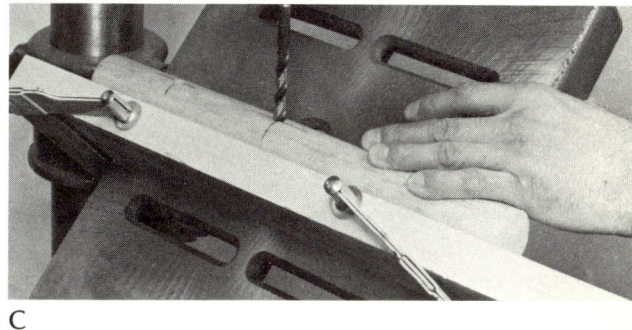

Fig. 3-23: Various applications of the V-block when drilling holes in circular work.

Round stock, as well as small odd-shaped pieces, can also be held in a clamp (Figs. 3-24A and B) or in a vise fastened to the table (Fig. 3-24C). On the workpiece, draw the center line of the hole on one end. Next, securely position the workpiece in the clamp or vise so that the center line of the bit to be used is aligned vertically and exactly in line with the drawn center line on the workpiece. Reposition the

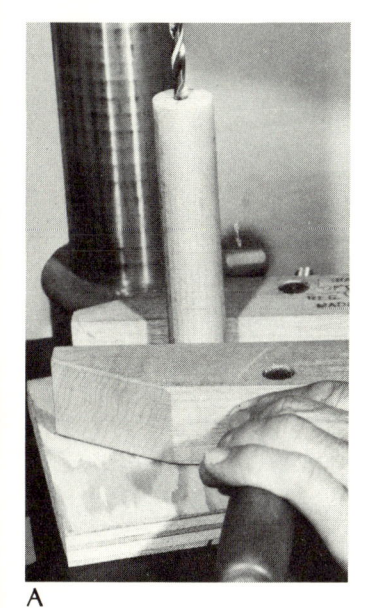

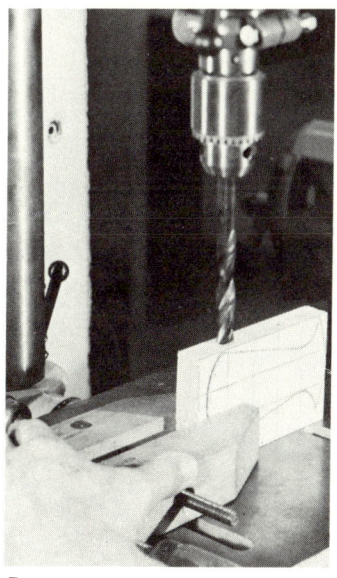

Fig. 3-24: Clamps and vises can be used to hold round and odd-shaped objects.

53

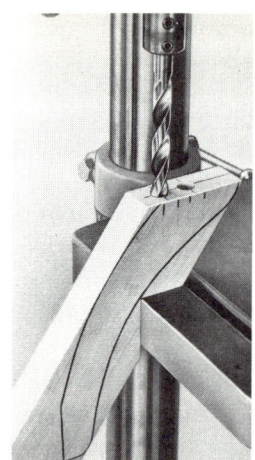

Fig. 3-25: Special jigs or wood block arrangements are sometimes necessary to drill odd-shapes.

clamp or vise, now with the work firmly clamped in its grasp, under the bit and bore the hole. Many occasions arise when it is necessary to devise special methods of adding wood blocking, such as when drilling odd-shaped pieces, to prevent the wood from spintering when boring (Fig. 2-25). Other odd shapes require special jigs.

Horizontal Boring. As stated in Chapter 1 and described in Chapter 2, the head of the radial type drill press swivels 360 degrees around its column and tilts more than 90 degrees right or left. Among the many

Fig. 3-26: The radial drill press as a horizontal boring machine.

Fig. 3-27: The horizontal boring operation has several applications.

advantages of this type of drill press is the ability to do horizontal drilling.

To use the radial drill press as a horizontal boring machine, a simple jig (Fig. 3-26) is needed to raise the material above the surface of the table top. The jig can be clamped to the table with C-clamps as shown. A high fence guide may be added to the jig for certain boring operations.

Additional applications of horizontal boring are shown in Fig. 3-27. Most operations that can be done in the vertical position can also be done in the horizontal one. The advantage of horizontal boring is being able to use workpieces of unlimited length.

BORING SCREW HOLES

Wood screws go in easier and hold better if properly sized holes are drilled for them. Usually there is a "lead" hole which allows the screw point to penetrate easily and lets the threads get a good grip without splitting the wood. There is also a "body" hole which provides clearance for the unthreaded screw shank. Two operations are generally necessary to bore most screw holes. The first hole to be drilled is the body hole. As shown in Fig. 3-28, this should be about the same size as the body of the screw. The smaller or lead hole is then drilled to take the threaded portion of the screw. The lead hole is about 70 percent of the screw size for softwood—90 percent for hardwood. The chart here should be consulted for the properly sized drill to use for various screws. When small screws are used in soft wood, the body hole and lead holes are unnecessary. A starting hole made with an awl is usually sufficient.

DRILL SIZES FOR SLOTTED WOOD SCREWS

Screw Size	BODY HOLE SIZE Nearest Fractional Size Drill	Accurate Drill Size No. or Letter	LEAD HOLE SIZE HARDWOODS Nearest Fractional Size Drill	Accurate Drill Size	SOFTWOODS Nearest Fractional Size Drill	Accurate Drill Size	Bit Size in 16ths for Counter-bore
0	1/16	52	1/32	70			
1	5/64	47	1/32	66	1/32	71	
2	3/32	42	3/64	56	1/32	65	3
3	7/64	37	1/16	54	3/64	58	4
4	7/64	32	1/16	52	3/64	55	4
5	1/8	30	5/64	49	1/16	53	4
6	9/64	27	5/64	47	1/16	52	5
7	5/32	22	3/32	44	1/16	51	5
8	11/64	18	3/32	40	5/64	48	6
9	3/16	14	7/64	37	5/64	45	6
10	3/16	10	7/64	33	3/32	43	6
11	13/64	4	1/8	31	3/32	40	7
12	7/32	2	1/8	30	7/64	38	8
14	1/4	D	9/64	25	7/64	32	8
16	17/64	I	5/32	18	9/64	29	9
18	19/64	N	3/16	13	9/64	26	10
20	21/64	P	13/64	4	11/64	19	11
24	3/8	V	7/32	1	3/16	15	12

Lead holes are seldom used for Nos. 0 and 1 gauge screws. In soft wood, lead holes are unnecessary for gauges less than No. 6.

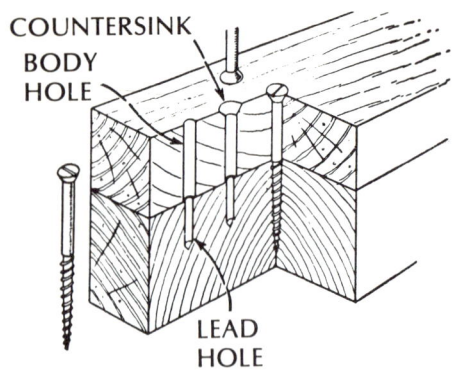

Fig. 3-28: Method of boring a screw hole.

Countersinking and Counterboring. A third operation often performed when boring screw holes is countersinking or counterboring. A countersink (Fig. 3-29A) is used when flat-head screws must be flush with the wood surface. A counterbore (Fig. 3-29B), on the other hand, is employed when the screw is set below the surface and then concealed with a plug.

Countersinking can be done with a regular countersink bit (Fig. 3-30A), or, a twist drill ground to an 82-degree angle can also be used. For production work, a combined drill and countersink, usually called a profile bit (Fig. 3-30B), is generally employed. The body and/or lead holes are bored first and then the countersink is made. When countersinking a series of holes, it is best to set the feed stop for a uniform depth. Center each hole by lowering the countersink into it before starting the drill press. On hardwoods, countersink to the full depth of the screwhead. On softwoods,

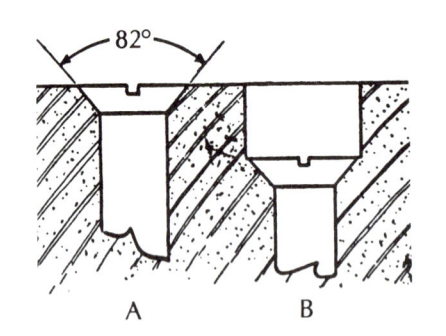

Fig. 3-29: (A) Countersink and (B) counterbore.

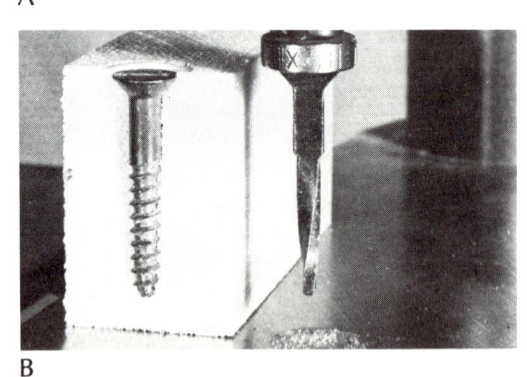

Fig. 3-30: A countersink bit (A) and a profile bit (B) at work.

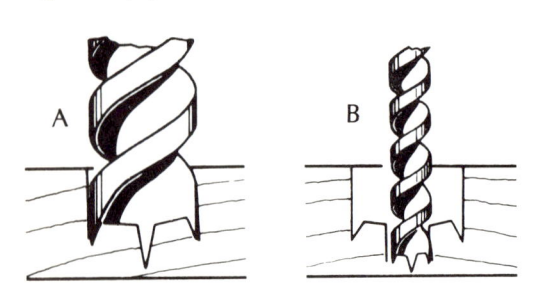

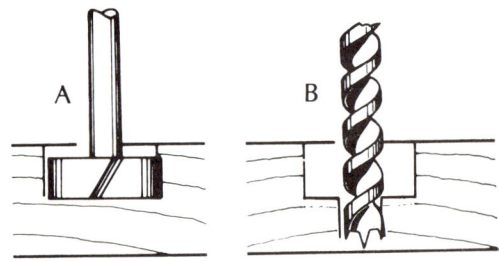

Fig. 3-31: When counterboring always use the large bit (A) first and then the smaller one (B).

the countersink should not be to the full depth. The screw will be pulled flush as it is driven in.

Counterboring can be done readily by using different sizes of bits, one for the counterboring and the others for the body and lead holes. When working with bits in this manner, the larger hole should be drilled first (Fig. 3-31), because it is usually more difficult to locate the larger drill in the hole center. Special counterboring bits which perform all three operations at once are also available (see page 9). Details on how to make plugs for counterbore holes are given later in this chapter.

ANGULAR BORING

There are three kinds of off-vertical or angular holes that can be done on a drill press: (1) simple angle, (2) equal compound angle, and (3) unequal compound angle. Each type of these angular holes can best be illustrated by considering the position of a table leg. For instance, at a simple angle, the leg's tilt would be in only one direction and it would be obvious when viewed from one side. With an equal compound angle, the leg would be off the vertical plane the same amount in two directions. Therefore, the angle would be the same whether the leg was viewed from the front or the side. The unequal compound angle also has the leg tilted in two directions, but more in one direction than the other. Therefore, the angle viewed from the front would be different than the angle viewed from the side.

Simple Angles. There are several methods of boring simple angles. If you have a tilting work table accessory, the simple angle can be done by tilting the table. The easiest way to set the table tilt is to use a table saw miter gauge, set at the required angle. Then sight across the arm of the gauge to the drill press column (Fig. 3-32A). Another way is to use any kind of set bevel, as shown in Fig. 3-32B. When the tilt is not excessive, no special provi-

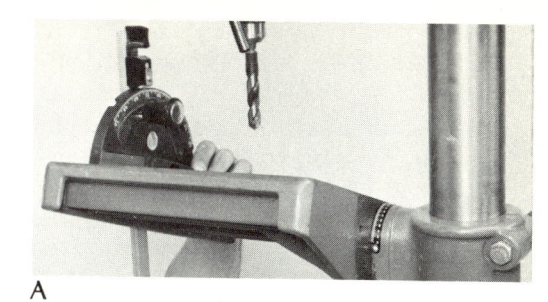

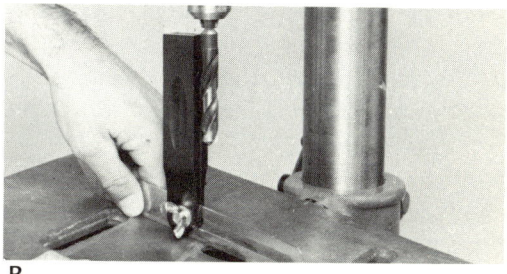

Fig. 3-32: Tilt settings can be made by sighting across a miter gauge or using a set bevel.

sions need be made to hold the work. It is simply placed on the table as it would be for ordinary drilling. The work must be square with the table within reasonable accuracy. Pencil lines on the table are an aid (Fig. 3-33), but setting by eye is normally accurate enough.

If the work size does not permit, or if you do not have a tilting work table, a wooden riser block placed under one edge of the

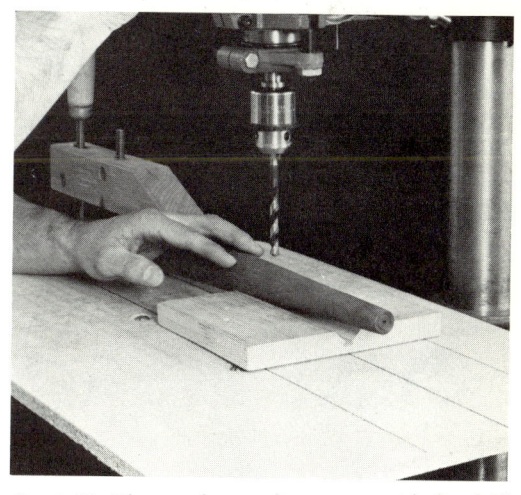

Fig. 3-33: The work must be square with the drill table.

57

work will permit simple angle boring (Fig. 3-34). Size these riser blocks to give yourself the simple angle desired. The auxiliary drill press tilting table described in Chapter 1 can also be used to bore simple angle holes (Fig. 7-35). In fact, this auxiliary tilting table can be used for most angular boring operations described in this chapter in the same manner as a standard tilting work table. To set the angle of the auxiliary table, use either a set bevel (Fig.3-36A) or a protractor. Small work can also be set at an angle and held in a clamp while boring (Fig. 3-36B).

With a radial drill press, the head can be tilted or swiveled right or left to the desired number of degrees (Fig. 3-37). Actually, any angle between 0 and 90 degrees can be bored with a radial machine. The method of setting the *head* is basically the same as tilting the table on the standard drill press, except the angle can be read directly off the scale (Fig. 3-38).

Equal Compound Angles. The forming of equal compound angles, as previously mentioned, involves a two-way tilt with both tilts equal, such as that required for the legs and corner posts of various

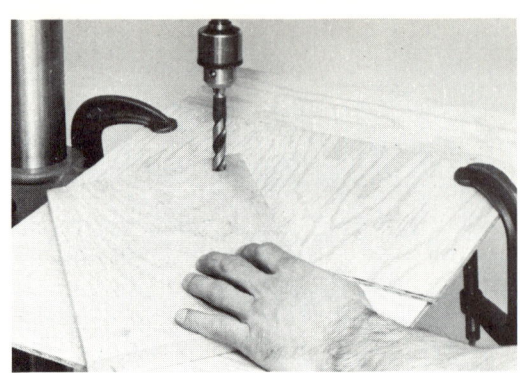

Fig. 3-34: A riser block along one edge will permit simple angle boring.

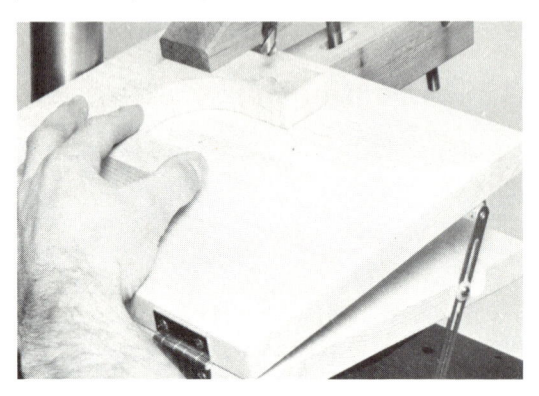

Fig. 3-35: Using an auxiliary tilt table to make a simple compound angle.

A

B

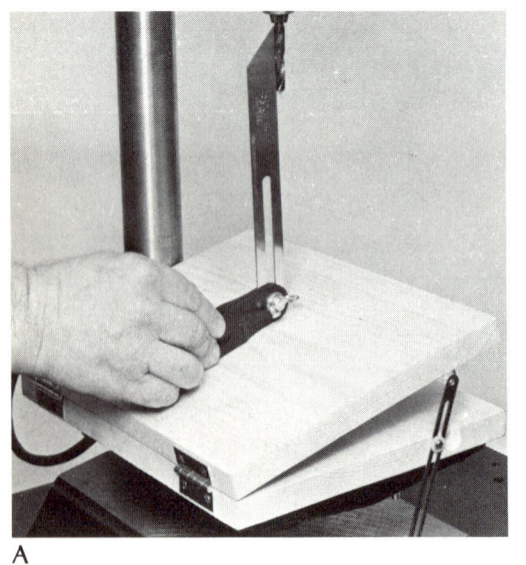

Fig. 3-36: Setting an auxiliary tilt table with a set bevel (A) and using a clamp to hold small work at an angle (B).

furniture work. The drilling line for such work is always at 45 degrees to the sides of the work, as shown in Fig. 3-39. To convert an equal two-way tilt to a single angle on a 45-degree drilling line, consult the chart here. For example, suppose the legs of a furniture piece are to be tilted 20 degrees as seen from the front and 20 degrees as seen from the end. The chart shows that a drill table tilt of 27 degrees will make the combined angles. (Riser blocks can also be used for equal compound angle hole boring.) The boring itself must be done on the 45-degree line. A simple setup for this is to use a V-block positioned square with the drill table. To use this setup for oval work, the holes should be bored before the oval shape is cut.

EQUAL COMPOUND ANGLES

Work Angle	Table Tilt	Work Angle	Table Tilt
2°	2 3/4°	12°	16 3/4°
3°	4 1/4°	13°	18 1/4°
4°	5 1/2°	14°	19 1/2°
5°	7°	15°	21°
6°	8 1/4°	17 1/2°	24 1/4°
7°	9 1/2°	20°	27°
8°	11°	22 1/2°	30 1/4°
9°	12 1/2°	25°	33°
10°	13 3/4°	27 1/2°	36°
11°	15 1/4°	30°	39°

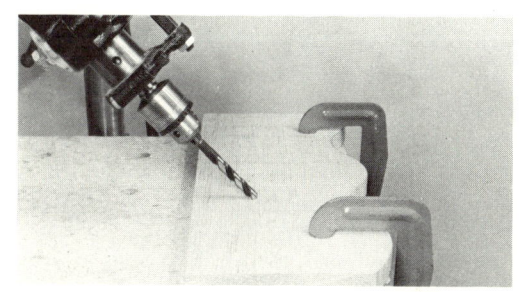

Fig. 3-37: Boring a simple angle with a radial drill press.

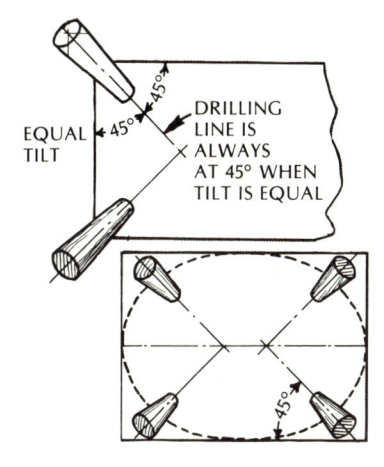

Fig. 3-39: Laying out an equal compound angle.

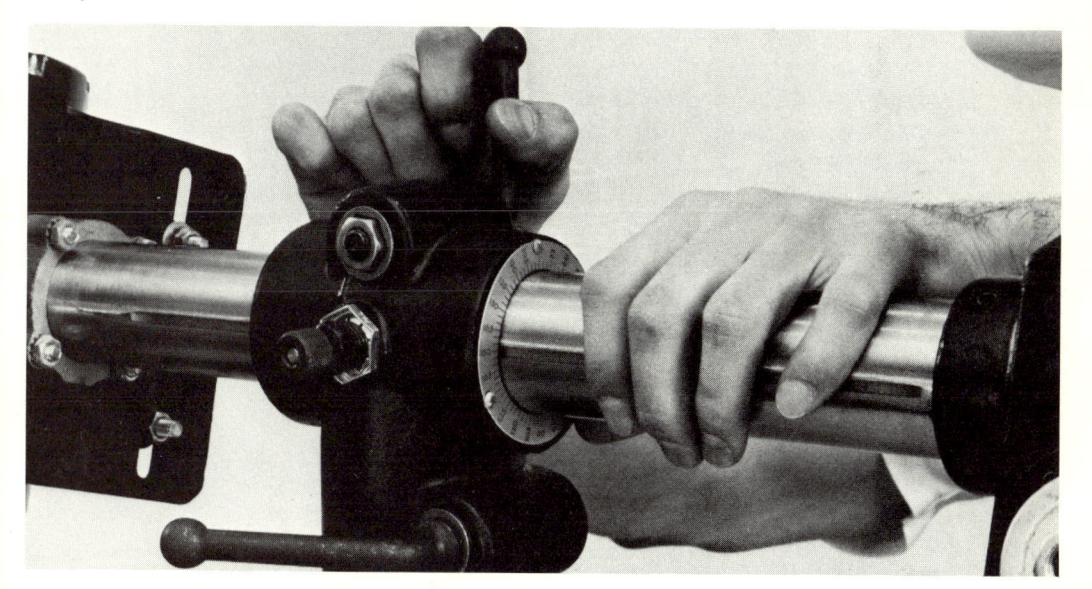

Fig. 3-38: Setting the radial drill press head.

When using a radial drill press, use the same table tilt from the chart here, for obtaining the number of degrees the drill head should be tilted to bore the equal compound angle (Fig. 3-40).

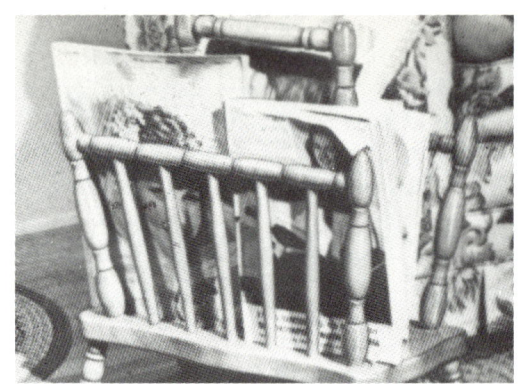

Fig. 3-41: A spindle magazine rack.

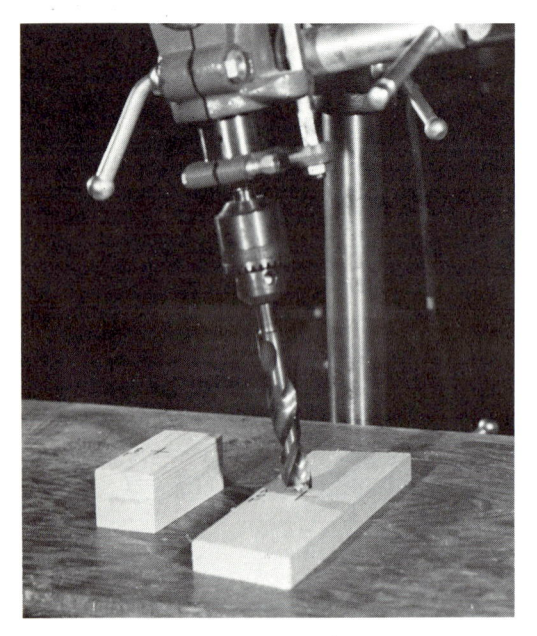

Fig. 3-40: Boring a compound angle on a radial drill press.

Unequal Compound Angles. Forming unequal compound angles means using a two-way tilt of different angles. For example, suppose you wished to set the spindles of the magazine rack at a uniform tilt of 8 degrees when viewed from the end, and 4, 6, and 8 degrees (the corner post) as viewed from the front. The simplest way to bore these angles with a standard drill is to use a direct two-way tilt on the work. In the example given, the drill table tilt or radial drill press head is used for the 4, 6, or 8-degree setting, while the uniform 8-degree setting is obtained with an 8-degree wedge or a riser. The work must be square with the drill press table. Figure 3-42 shows how to drill an unequal compound angle hole such as is needed for the spindle next to a corner post. This requires a 6-degree tilt of the drill table or radial head to set the angle as seen from the front. The uniform 8 degrees at which all spindles are tilted, as seen from the end of the magazine rack, is obtained by means of a riser block which can be seen under the work. The corner post is an equal compound angle of 8 degrees and could be drilled by using the conversion angle of 11 degrees on a 45-degree drilling line. However, since the setup for other holes is made with a two-way tilt, the corner post holes are worked in the same fashion.

A drawing board method of laying out a compound angle is shown in Figs. 3-43 and 3-44. Make a bottom view of one corner of the work, and on it locate the leg centers

Fig. 3-42: Unequal compound angle can be set with two-way tilt, using the table tilt for one tilt and riser block for the other.

and the leg center line, A and B (Fig. 3-43). From B, erect a perpendicular equal to the height of the work up to the underside of the top. This will establish point C. Connect CA. CA is the true length of the leg, and angle C is the drilling angle. To take off the angle, set an adjustable square at 90 degrees and then add angle C. Use the square set at this position to set the drill press table (Fig. 3-44). When drilling the hole, the leg center line must be a continuation of the drill center line as viewed from the side. In other words, the line on which you are drilling must be square with the drill press table or auxiliary tilting table.

When angle holes are required in the ends of table legs for dowels, it is often practical to make a jig to hold the work at the required angle (Fig. 3-45). However, most projects of this nature can be worked more easily by first sawing the end of the leg at the required angle. This establishes a small but definite working surface which can be presented squarely to the drill by using the handscrew setup as shown on page 65.

In all angular drilling, it is important to keep in mind that the side of the bit frequently makes contact with the stock before the point does. This condition could cause the bit to wander, unless the bit is fed very slowly, until it is firmly in the work. Sometimes on extreme angles, it is a good idea to employ a leveling block which will provide a flat surface for the bit to enter.

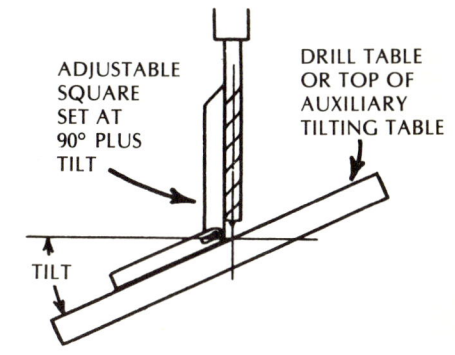

Fig. 3-44: Setting the tilt of the drill table or auxiliary tilting table.

Fig. 3-43: The drawing board method of laying out a compound angle.

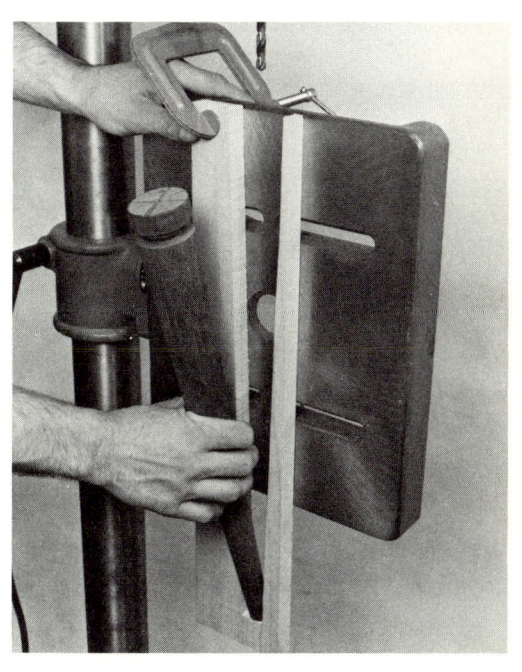

Fig. 3-45: Holding jigs are sometimes useful for angle boring.

61

Fig. 3-46: Typical use of a pocket hole.

Fig. 3-47: Block beveled at 15 degrees, makes a good set-up for drilling pocket holes.

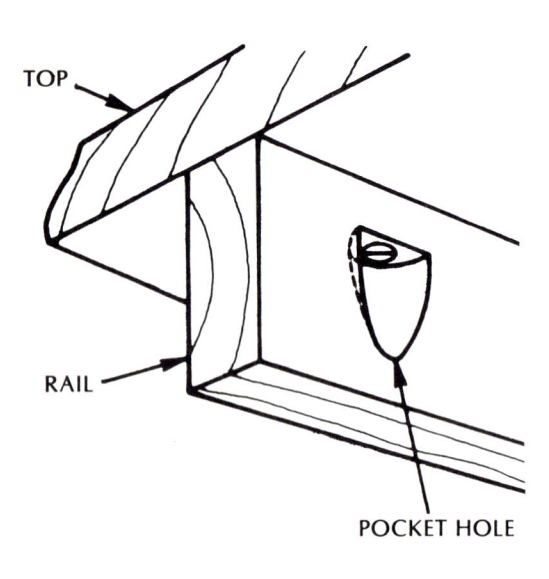

Pocket Holes. Pocket holes are used to fasten table tops (Fig. 3-46) and other work. This type of drilling requires a simple one-way tilt, best obtained by using a heavy block of wood beveled at about 15 degrees. A hole bored through the edge of the bevel will center properly at the top edge of the rail (Fig. 3-47). The hole in the beveled block supports the drill and prevents run-out. The same setup can be used for a corner pocket hole (Fig. 3-48), the only difference being that the work must be turned to a 45-degree position to establish the required drilling line. The work can be held freehand in this position because exact accuracy is not usually required.

DOWELS

Hardwood dowels can be purchased in a variety of sizes or can be made in the shop with the use of a plug cutter. The most useful size is the 3/8-inch diameter, and the most common lengths are 1 3/4 and 2 1/2 inches. Dowels should fit snugly but should not be forcibly driven in place with a hammer. The combined length of the dowel holes should be a little more than the dowel length to allow the proper assembly and to provide room for excess glue. Frequently, spiral cut dowels are used because they allow the glue to flow freely and the air to escape from the dowel hole. (Fig. 3-49).

Locating Dowel Holes. Several methods of locating dowel holes are shown on page 38. Undoubtedly, the system most used is with dowel pops or pins. When using dowel pops to locate holes for a joint, the holes should be drilled in the half of the joint which is most difficult to fit. For example, in a plain corner joint, the holes should be drilled in the end-grain piece (Fig. 3-50). The idea is that the holes for the dowel pops or pins need not be located exactly. A slight drift in the end grain does little harm.

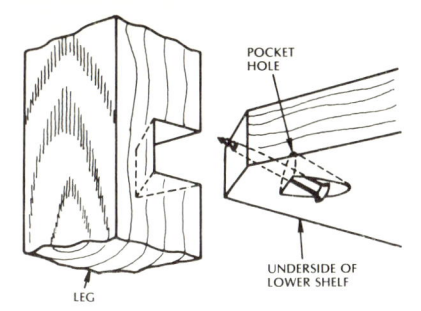

Fig. 3-48: Another use of a pocket hole.

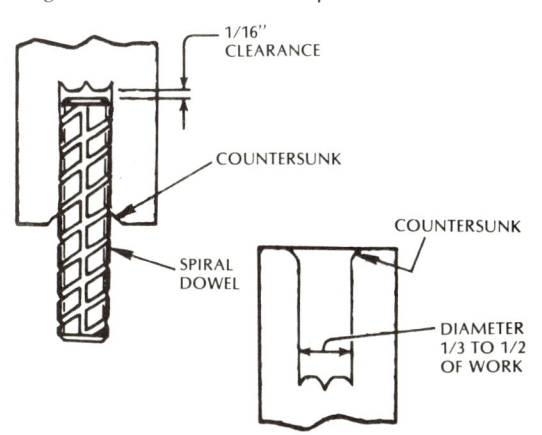

Fig. 3-49: Standard dowel construction.

Fig. 3-50: Dowel pops or pins offer a simple and accurate method of locating dowel holes.

Fig. 3-51: One method of locating dowel holes on duplicate pieces.

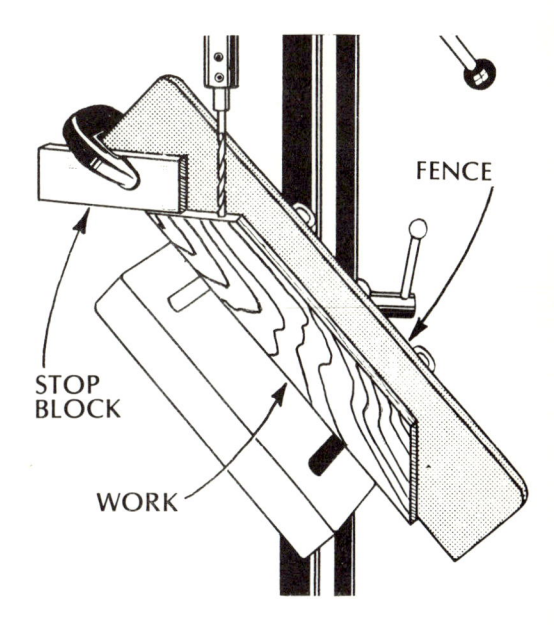

Fig. 3-52: The stop block automatically locates the position for the first dowel hole.

After the holes have been drilled and the dowel pops fitted, the two members of the joint must be brought together in exact alignment. In the plain corner joint, this is done by guiding the two members along a straight edge, such as the fence of the table saw (Fig. 3-50). Care should be taken when aligning, that the finished or face side of any joint should be placed on the flat table surface. The second set of holes must be drilled exactly as indicated by the marks made by the dowel pops or pins. Accuracy is aided by the fact that the second of the two pieces is the easier to drill.

Mechanical Spacing. When similar pieces are to be drilled for dowels, mechanical spacing is faster and more accurate than the use of dowel pops or pins. A typical example is the edge-to-edge joint (Fig. 3-51) where any number of pieces can be drilled accurately for dowels by using a wood fence and stop pins.

Another example is the miter joint. One method of doing this job is shown in Figs. 3-52 and 3-53. The drill table is tilted 45 degrees, making the miter cut horizontal. A fence is used to accurately center the holes, with respect to the thickness of the work. A stop block clamped to the fence locates the position of the first hole which is being drilled in Fig. 3-51. No bottom support is used. The work is simply held tightly against the fence. The depth stop should be set to the proper depth.

After the first hole has been drilled in all pieces, a suitable spacer block is placed below the block clamped to the fence, and the work is pushed up until it contacts this block (Fig. 3-53). The spacer is then laid aside, holding the work firmly against the fence so it will not slip, and the second dowel hole is drilled.

Where a tilting work table is not available, the auxiliary table can be substituted and can be used to bore dowel holes for a miter joint (Fig. 3-54) in the same manner as just described. With a radial drill press, miter joint dowel holes are drilled with the head tilted 45 degrees as shown in Fig. 3-55. Sometimes dowel holes must be located in the end grain of a workpiece that is cut at an angle. To do this, adjust the table to this angle by tilting the table (Fig. 3-56A) or the head (Fig. 3-56B) so the hole is at right angles to the end grain.

How the work is held when drilling a dowel hole or holes in a circular segment is shown in Fig. 3-57. The table or head (in the case of the radial type) is tilted and a suitable support to carry the work is clamped into place. A stop block is nailed to the support to stop the work at the required distance. Each segment of the

Fig. 3-54: Boring a dowel hole for a miter joint on the auxiliary table.

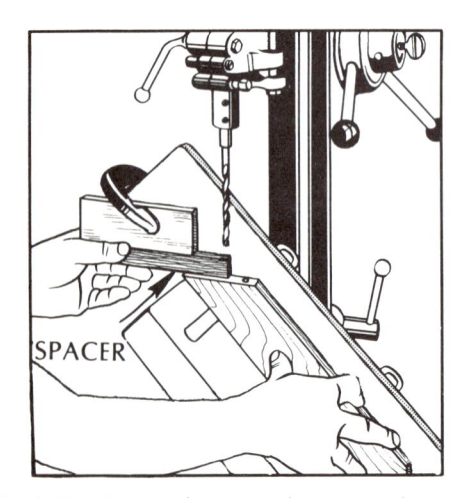

Fig. 3-53: A wood spacer between the stop block and the work locates the position for the second hole.

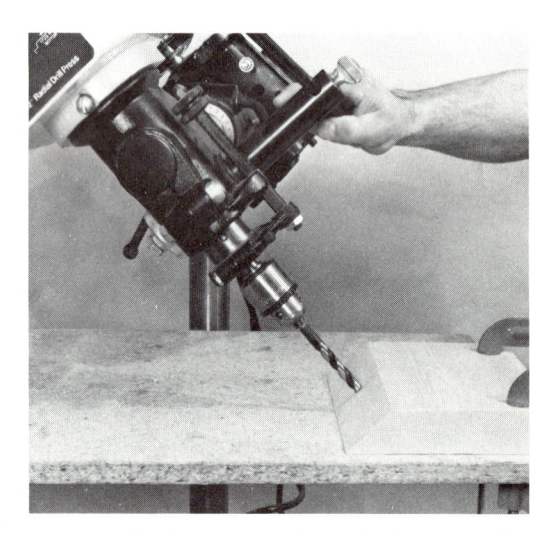

Fig. 3-55: Boring a miter dowel hole on a radial drill press.

Fig. 3-56: Boring a hole in end grain with the table (A) in a vertical position or head (B) tilted in the horizontal position.

wheel is drilled with the same setup, insuring the proper fitting of the dowel holes in each piece.

Jigs or methods of working, similar to those described, can generally be made for any piece of work which is to be fastened together with dowels. Jigs of this nature are particularly advantageous in production work. Considerable time is saved and better work is accomplished by locating the holes mechanically instead of marking each individual piece with dowel pops or similar methods.

When making corner joints with dowels, one of the two members will usually require end drilling. A good device for presenting the work squarely to the drill consists of an ordinary handscrew clamp with a guide strip screwed to one of the jaws, as shown in Fig. 3-58. When the work is fitted in the handscrew with the end surface butted against the guide strip, it will be square with the drill when the handscrew is held on the drill table (Fig. 3-59). This setup can be used for miters, or for any kind of "short surface" dowel joint, because the joining surfaces must always be square with the drill regardless of the type of joint.

A *depth stop*, which is fitted directly over the drill, is shown in Fig. 3-60. This is a faster and more convenient method of working than setting the depth stop on the drill

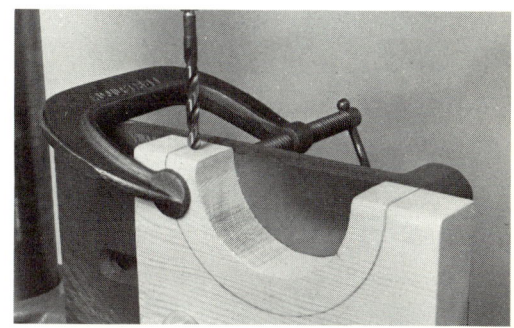

Fig. 3-57: Boring in circular work.

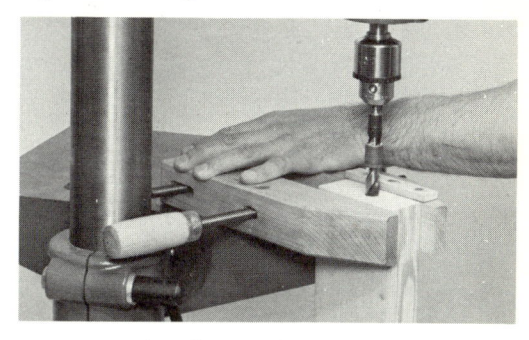

Fig. 3-58: A handscrew clamp with a stop guide screwed to one jaw makes a good device for holding work for end drilling.

Fig. 3-59: The handscrew clamp guide.

65

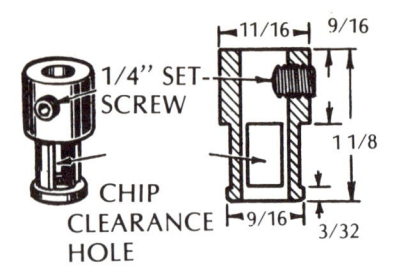

Fig. 3-60: Depth stop assures equal and correct depth of dowel holes.

press. The stop shown is designed for a 3/8-inch machine spur bit. Other sizes can be made by changing diameters as needed.

Typical dowel joints are shown in Fig. 3-61. To gain strength when doweling a rail to a table leg, the dowels should be either offset or mitered, as shown in the examples at the left, in Fig. 3-61. Offsetting is by far the stronger method but it is not always possible on narrow work. Often, good use can be made of the simpler through-doweling, as shown in the center examples. This kind of joint is easy to fit since the holes are drilled through both parts of the work after assembly. The dowel holes are sometimes slanted slightly to make the joint less likely to pull apart.

PLUG-CUTTING SPEEDS

Plug Size Inch	Suggested Spindle Speed — RPM	
	Softwoods	Hardwoods
3/8	2400	1250
1/2	1250	1250
5/8	1250	1250
3/4	1250	700
1	700	700

CUTTING DOWELS AND PLUGS.

Both dowels and plugs can be cut from stock with plug cutters. The plug-cutter speed depends on its size and the type of wood it is cutting.

While dowels can be purchased from hardware stores, lumber yards, and home centers, they can be made easily in the workshop. They are generally cut in the end grain of the wood (Fig. 3-62) so that the dowel will have a strong cross-section. Free the dowels from the stock by making a saw

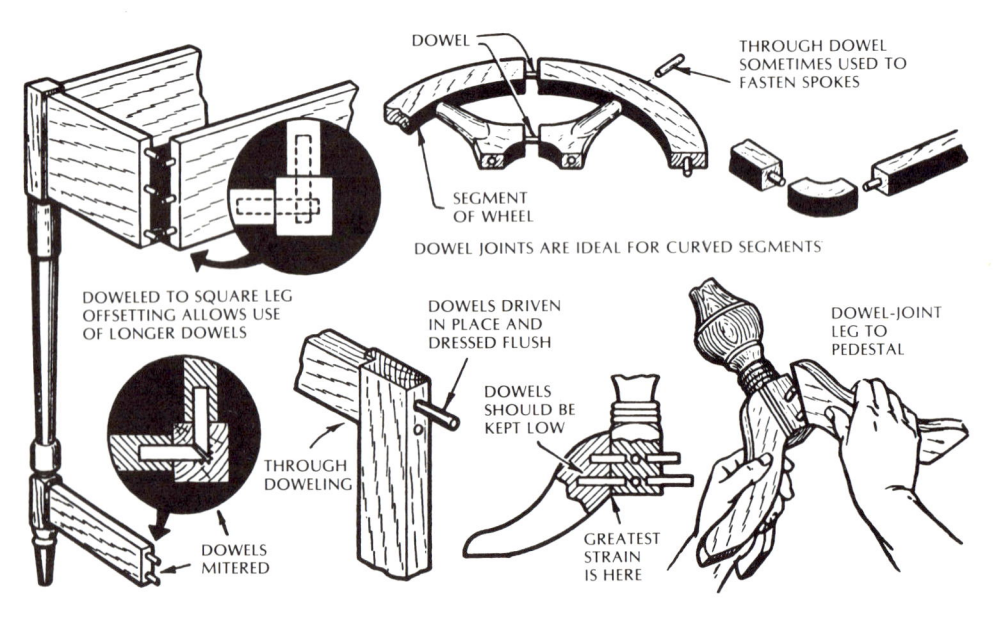

Fig. 3-61: Typical dowel joints.

Fig. 3-62: Cutting dowels on the drill press.

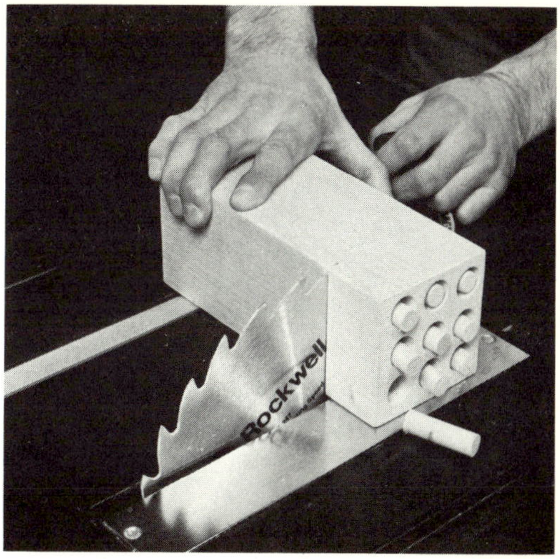

Fig. 3-63: Free the dowels by making a saw cut. (Note: Blade guard was removed for picture clarity; cut should be made with all saw safety devices in place.)

cut as shown in Fig. 3-63. The length of dowels can be determined by this rip cut. But, never make this cut so that the loose dowels will be between the rip fence and the blade.

Plugs are usually cut from stock equal to the diameter of the plug, that is, a 3/8-inch plug cutter uses 3/8-inch stock. Since the most common use of plugs is to conceal screw or bolt holes, use crossgrain plugs (Fig. 3-64) cut from the same material as the project. You can match the grain direction, and even the wood tone if you take a little care.

Pointers in the fitting of plugs are shown in Fig. 3-65. A suitable dowel rod (Fig. 3-65B) can be used to press plugs flush. As in other counterboring jobs, the plug hole is drilled first so that the drill will leave a center mark to locate the smaller drill used for the screw hole. Plugs should be fitted in place with the grain (Fig. 3-65A), using

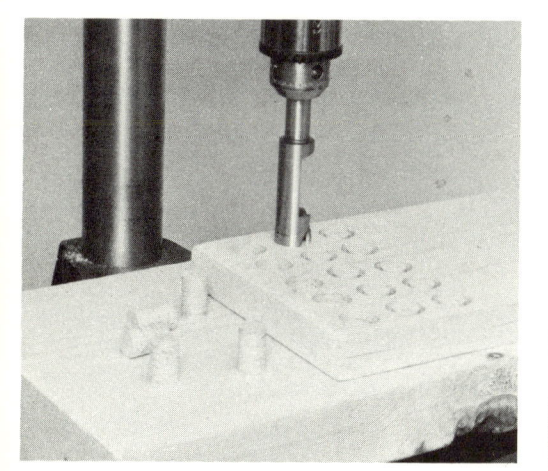

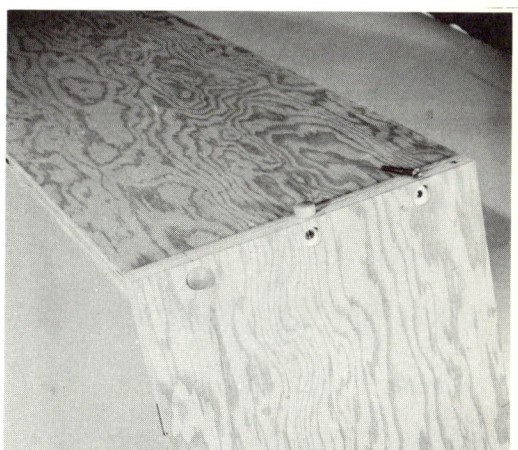

Fig. 3-64: Cutting plugs (left) and how they fit (right).

shellac or glue. When dry, the surplus shellac or glue is chiseled off slightly above the surface (Fig. 3-65C) and the whole surface is sanded. Common wood plug faults are shown in Fig. 3-65D.

Round Tenons. Round tenons can be cut directly on the end of any piece of work by using a plug cutter, as shown in Fig. 3-66A. The work is first cut on a table saw or band saw to match the diameter of the round tenon. The plug cutter is then used to meet these kerfs, leaving a perfectly round tenon (Fig. 3-66B). This setup is especially good for small work which would be weakened by fitting a dowel in the ordinary manner. By using a jig such as shown in Fig. 3-67, it is possible to cut integral tenons on round stock.

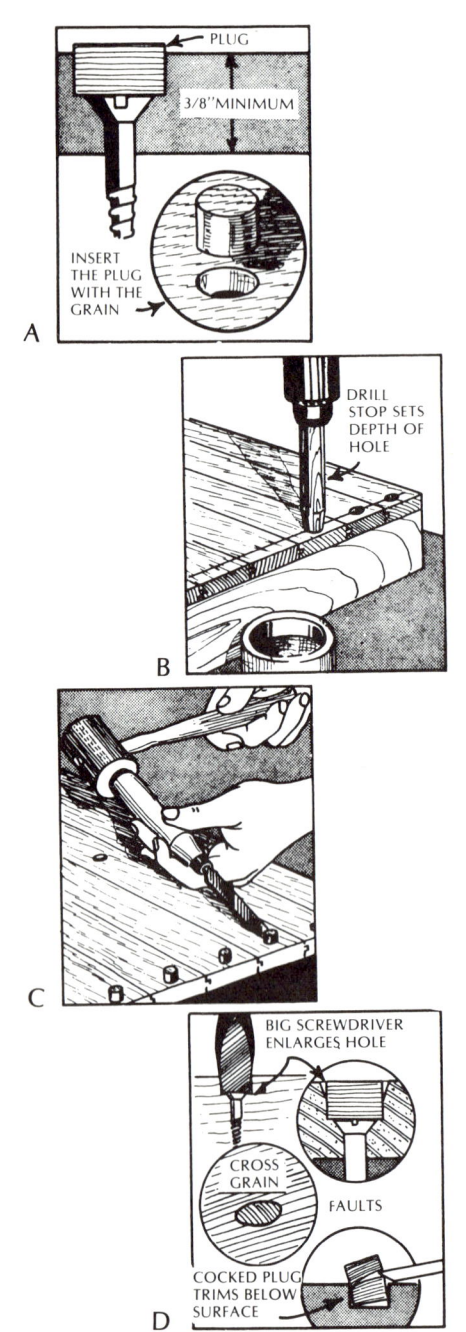

Fig. 3-65: Pointers in fitting plugs.

Fig. 3-66: How to cut an integral tenon on square stock.

Fig. 3-67: How to cut an integral tenon on round stock.

DECORATIVE BORING

Decorative works of art for the fronts of drawers and door pulls can be created on a drill press. A great deal of this work is done with a fly cutter that has a sloping bit rather than the normal vertical bit. When the tool is fed, the pilot drill makes a center hole while the blade makes a recess which is circular and sloping outward from the bottom. As shown in Fig. 3-68, for example, boring decorative concentric circles is done in successive steps with the cut diameter reduced for each. After boring, the outside of the part can be shaped as a circle, oval, square, and so on. With a fly cutter and a little imagination many interesting designs can be created. While this type of decorative work can be accomplished on almost any wood, stock such as birch is best.

Fig. 3-68: Boring decorative concentric circles.

VARIOUS JOBS

As mentioned earlier in the chapter, there are many boring operations that require special jigs or set-ups to hold the work. This is especially true when end boring. For instance, a hand-screw clamp and a wood fence (Fig. 3-69) is a simple way

Fig. 3-69: A simple way of holding work when end boring.

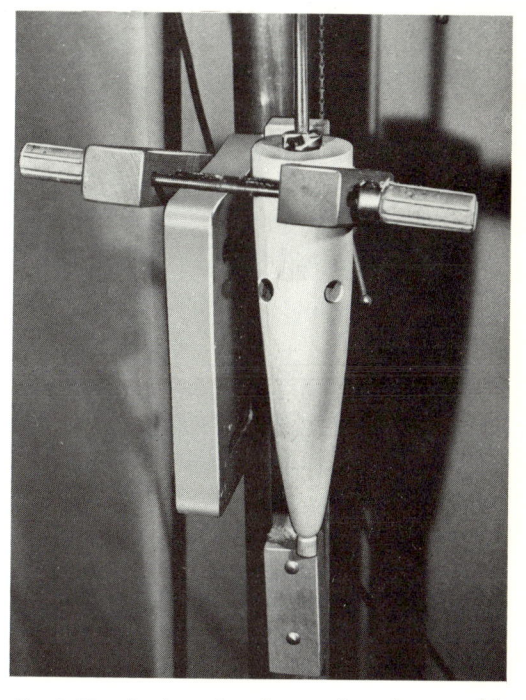

Fig. 3-70: Boring in the grain of an odd-shaped piece.

of holding work for end boring, while the odd-shaped piece illustrated in Fig. 3-70 utilizes a clamp, plus a fence with a wood block fastened to one end. The two-sided bevel boards shown in Fig. 3-71 are end bored in a special V-shaped jig.

The setup shown in Fig. 3-72 is ideal for long work. It is merely an arrangement of three pieces of wood, sawed carefully to give an exact right angle when placed on the drill press table or base. Work clamped in the jig will then be positioned properly for drilling.

Various types of adjustable fences can be used for end-boring in long work. A typical form of construction is shown in Fig. 3-73, designed for a 14-inch drill press. The column collar is clamped to the drill press column by means of a wing nut, while the fence itself is set at the required position to suit the thickness of the work. A strip of wood clamped to the fence makes a right angle stop. The bottom of the work is supported on the drill table, as shown in Fig. 3-74.

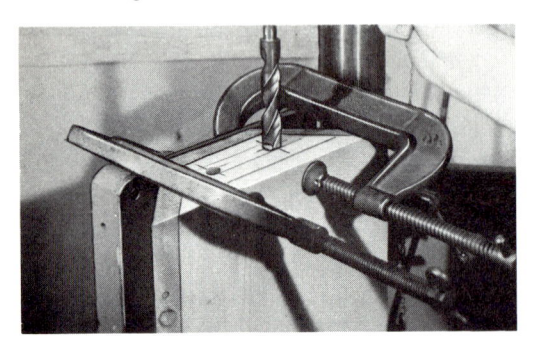

Fig. 3-71: Jigs sometimes must be especially made to hold work.

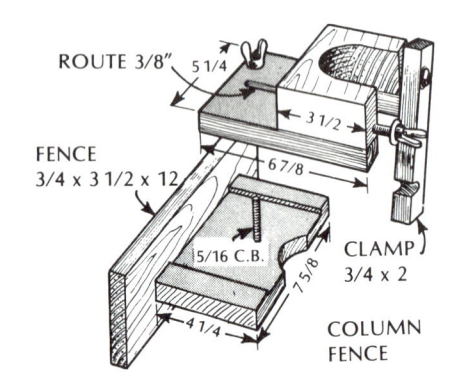

Fig. 3-73: Details of a column fence.

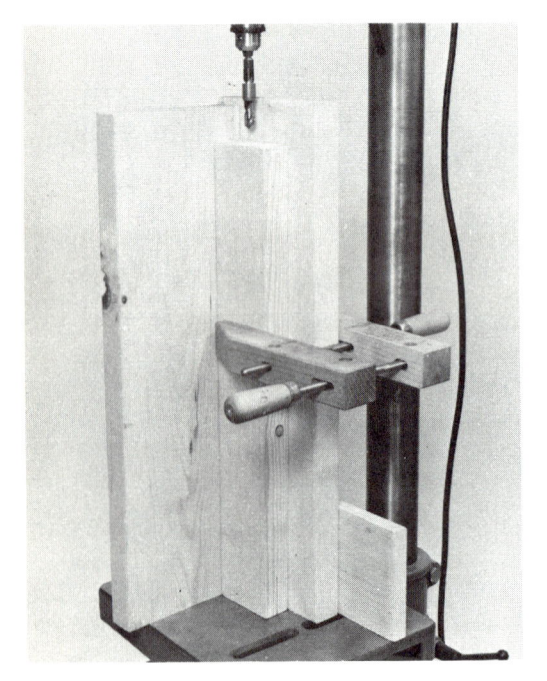

Fig. 3-72: A simple squaring block.

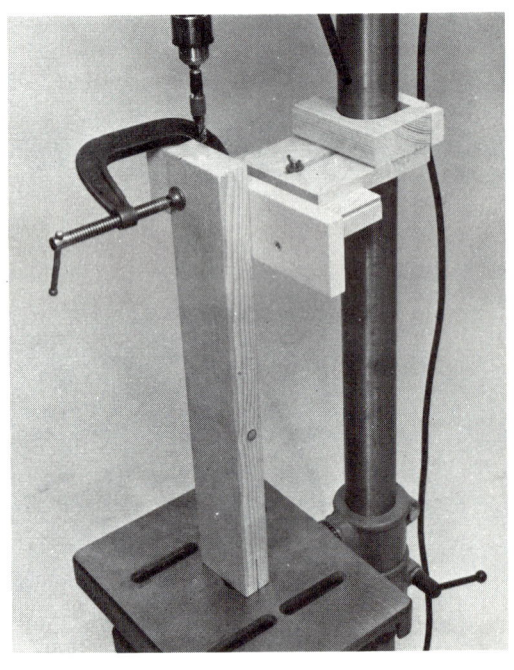

Fig. 3-74: A column fence in use.

Chapter 4

MORTISING ON THE DRILL PRESS

The mortise-and-tenon joint is one of the strongest and most durable in all woodworking. While the tenon is cut easily on a table saw, the mortise can be a problem unless a special mortising attachment is used in conjunction with the drill press. This attachment eliminates the tedious handwork required to cut the rectangular cavity needed for the tenon.

INSTALLING THE MORTISING ATTACHMENT

A typical mortising attachment is shown in Fig. 4-1. The mortising chisel is held stationary by means of a cast chisel holder attached to the drill press quill. The mortising bit or chisel can be held in a spindle with a standard chuck or a 1/2-inch hole. With the latter, a bushing is slipped over the end of the bit to adapt it to the hole. A fence, which is necessary to keep the mortise straight and accurate, is fastened to the drill press table. The fence carries a bracket on its top edge, and this bracket supports an adjustable hold-down. The bracket also carries hook rods or hold-ins which are used to keep the work against the fence.

Inserting the Chisel. The bit is slipped through the chisel from the cutting end; then the proper size bushing is placed on the shank of the bit and the assembly is slipped upward through the hole in the chisel holder. Both chisel and bushing should be pushed up until they butt against their respective shoulders. The chisel is secured by tightening setscrews. The bit is then secured, after first carefully adjusting it so that the spurs of the bit are at least 1/32 inch, but not more than 1/16 inch, away from the lower end of the chisel (Fig. 4-2).

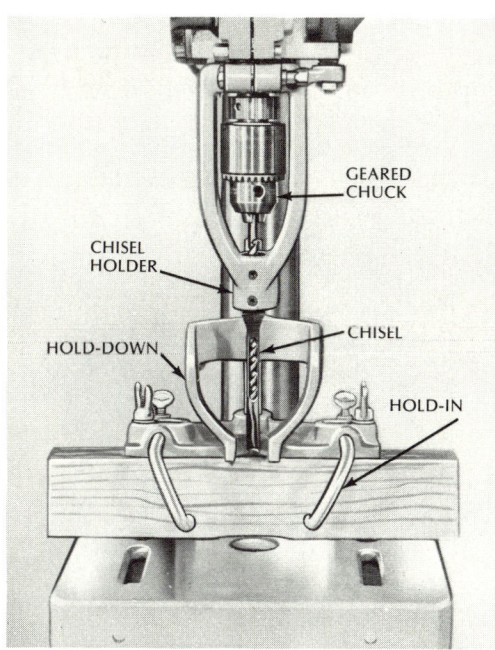

Fig. 4-1: The drill press converted for mortising.

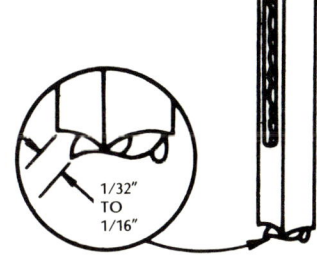

Fig. 4-2: Bit clearance is important to prevent chip clogging.

This is very important. Keep the clearance to a minimum but not too tight to create a friction between the bit and the chisel. Insufficient clearance will cause overheating and could result in damage to the cutter. On the other hand, the bit should not project too far, since the chips produced may be too large to pass through the chisel, clogging the tool.

If the bit appears to rub on the chisel, check the extension of the bit. If it is correct, loosen the chuck screws, turn the chisel a trifle inside the bushing and retighten. Tighten the setscrews in the spindle chuck so that they bear against the flat on the bushing.

A method for ensuring proper clearance is as follows. Insert the bit into the chisel with the spurs of the bit against the cutting edge of the chisel. Place both chisel and bit into the chisel holder and while maintaining contact between the bit and chisel end, position the shoulder of the chisel 1/32 inch below the bottom surface of the chisel holder. Tighten the chuck or set screws firmly on the bit. Then raise the chisel completely and lock.

Because the fence is used as a guide for the workpiece, the chisel must be squared to it. Place the head of a combination square against the fence and the blade against the side of the chisel (Fig. 4-3A). If the chisel is not square with the fence, the mortise cuts will not be square with the fence, as shown in Fig. 4-3B. The quill is lowered to advance the chisel to the proper depth (Fig. 4-3C), and the depth stop is set to this position. If the cut is to go through the work, use a scrap or base block to protect the table, chisel, and bit.

The hold-down and hold-ins are now adjusted so that the workpiece is held in place firmly, but free to slide along the table.

The final step is to adjust the speed of the mortising chisel bit. Generally, the larger the chisel, the slower the speed should be, especially when mortising hardwoods. The chart below lists suggested spindle speeds for mortising operations.

MORTISING SPEEDS

Mortise Size—Inch	Suggested Spindle Speed—RPM Softwood	Hardwood
1/4	2400	1250
5/16	2400	1250
3/8	2400	1250
1/2	1250	1250

These speeds apply for average conditions. Keep in mind that there are variations in both softwoods and hardwoods, even in boards sawed from the same tree. Also consideration must be given to whether you are cutting with the grain, across the grain, or into the end grain.

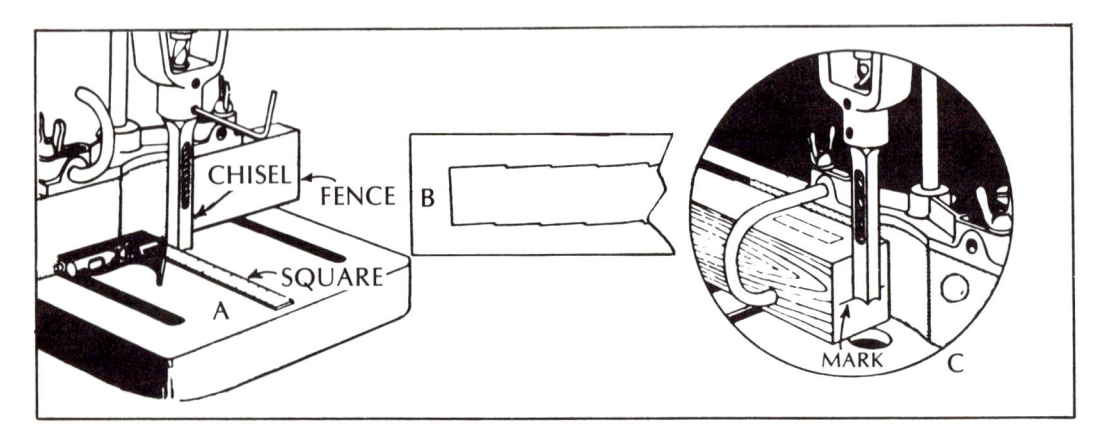

Fig. 4-3: The fence must be square and parallel with the chisel, otherwise the cuts will be staggered.

Fig. 4-4: Cutting a mortise.

CUTTING THE MORTISE

After the workpiece has been located on the table, the cut is started by bringing the chisel down at one end of the mortise (Fig. 4-4). Because all four sides of the chisel are cutting, the cut on the first stroke will naturally take a little more pressure than the following ones. Remember, avoid excessive feed pressure. Nevertheless, a light, feather touch will not work either because the chisel would be cutting only under quill-feed pressure. Actually, the spindle speed and rate of feed are usually considered correct when the waste chips move easily up the flutes of the bit and fall out through the escape slot in the chisel without blockage. Be sure to retract the chisel frequently, especially with hardwoods, to clear the chips and permit the tool to cool.

When boring blind mortises, the quill-feed stop-rod can be used to control the depth of the cut. To minimize splintering on through mortises, use a scrap block under the workpiece. Where the surface appearance of through mortises is important, it is usually best to use stock that is slightly thicker than is necessary for the part. Then, after the mortising operation is completed, a thin shaving cut on the table saw will remove any imperfections.

Mortises longer and/or wider than the chisel width require two or more cuts.

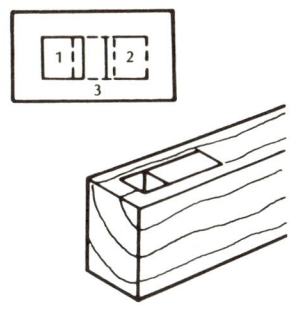

Fig. 4-5: Cutting a long mortise.

Make the first cut at one end or corner of the mortise outline. Once the first cut has been made to the depth desired, the work is moved sideways to the other end or corner and the cut is again made to full depth. After the end or corner cuts are completed, proceed to clean away the stock between by making overlapping cuts (about 3/4 of the full chisel width) to full depth as shown in Fig. 4-5. The work must

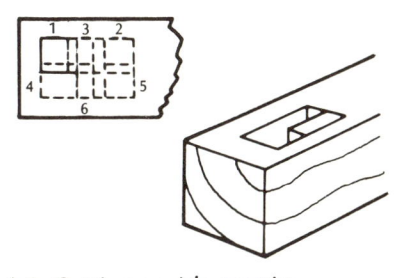

Fig. 4-6: Cutting a wide mortise.

be repositioned carefully for each new cut. Where the work demands a mortise wider than the chisel, set the fence at the proper distance and make a second series of cuts as shown in Fig. 4-6. Avoid narrow shoulders. The chisel will creep away and leave a tapered side. Many pieces of work split because tenons are forced into mortises with sloping sides. If necessary, use a smaller chisel.

As mentioned earlier, tenons are generally cut on a table saw. For more information, see *Getting the Most Out of Your Table Saw*. However, when cutting a tenon, remember not to size it so that it must be forced into the mortise. Also, remember not to cut it too long. There must be room for excess glue. Keep the tenon shorter by about 1/16-inch or chamfer its end.

BLIND-WEDGED TENON

MORTISE AND TENON WITH SPLINES

CONCEALED HAUNCHED TENON

TENON WITH LONG AND SHORT SHOULDERS

STUB TENON

MITERED TENON

HAUNCHED TENON

BARE FACED TENON

SIMPLE MORTISE AND TENON

THROUGH-WEDGED TENON

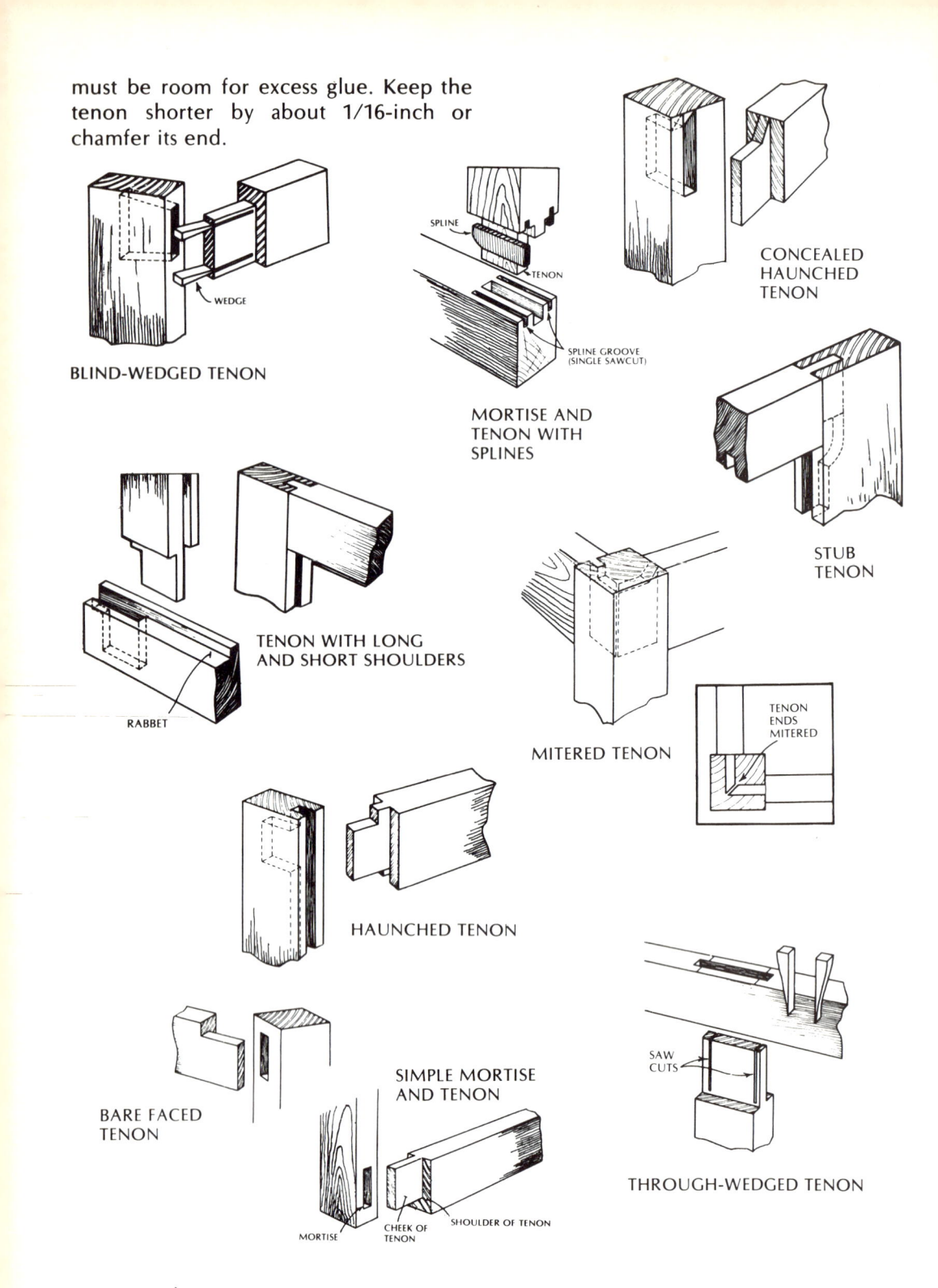

Fig. 4-7: Various mortise-and-tenon joints.

Various Mortise-and-Tenon Joints. There are many different variations of the simple mortise-and-tenon joint. Some of the more popular ones are shown in Fig. 4-7. Note that mortise-and-tenon joints are normally named after the type of tenon.

The *blind* or *simple tenon*, is the most common, and is used primarily in furniture framework.

The *mitered tenon*, frequently used in table construction, is used to secure the maximum length of tenon. Each joint is a simple mortise and tenon with the tenon end mitered at 45 degrees as shown. The two mortises meet at 90 degrees inside the vertical (leg) member. Mitering of the tenon ends allows for deeper tenons.

The *bare-faced tenon* has but one shoulder and is used when a tenoned piece is thinner than a mortised piece.

The *through-wedged* or *fox wedge tenon* is useful where both added strength and resistance to pulling apart is required. The two ends of the mortise are sloped outward to provide room for the wedges, which are about half the tenon in length.

The *blind-wedged tenon* is used in the same way as the through-wedged tenon, but can be employed in a location where the through-wedged cannot be used.

The *tenon with splines* can only be used when work is 1 3/8 inches or more in thickness. The splines guard against the work twisting out of line.

The *tenon with long and short shoulders* is used in a framework or sash where a rabbet is required.

The *stub tenon* is not a true mortise-and-tenon joint, but it is easily made and is useful for light framing.

The *tusk tenon* is used on some furniture styles as a decorative detail. The square hole in the tenon end is spaced so that the tusk key forces the rail against the shoulder of the tenon.

The *haunched tenon* is employed where added tenon strength is needed and where partial exposure on top is not objectionable.

The *concealed haunched tenon* (Fig. 4-8) gives the needed extra strength to the joint without showing a break at the end. It is used frequently on light table legs where a square haunch would weaken the construction. In making this type of joint, the tenon is cut first. The square shoulder cuts are run in first, after which the saw arbor is tilted about 30 degrees for the beveled haunch (Fig. 4-9A). The cheek cuts are then made (Fig. 4-9B) to complete the

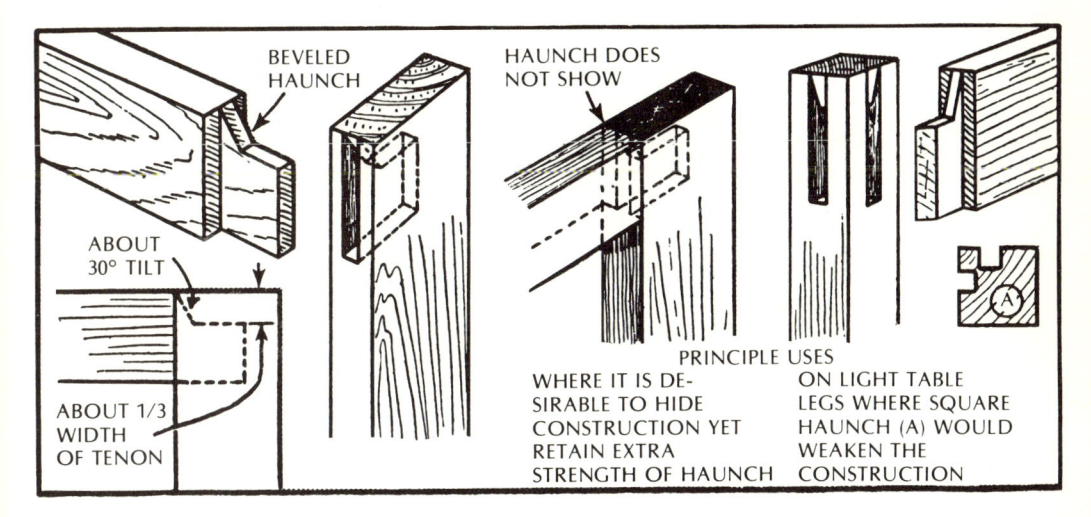

Fig. 4-8: The concealed haunched tenon.

A

C

B

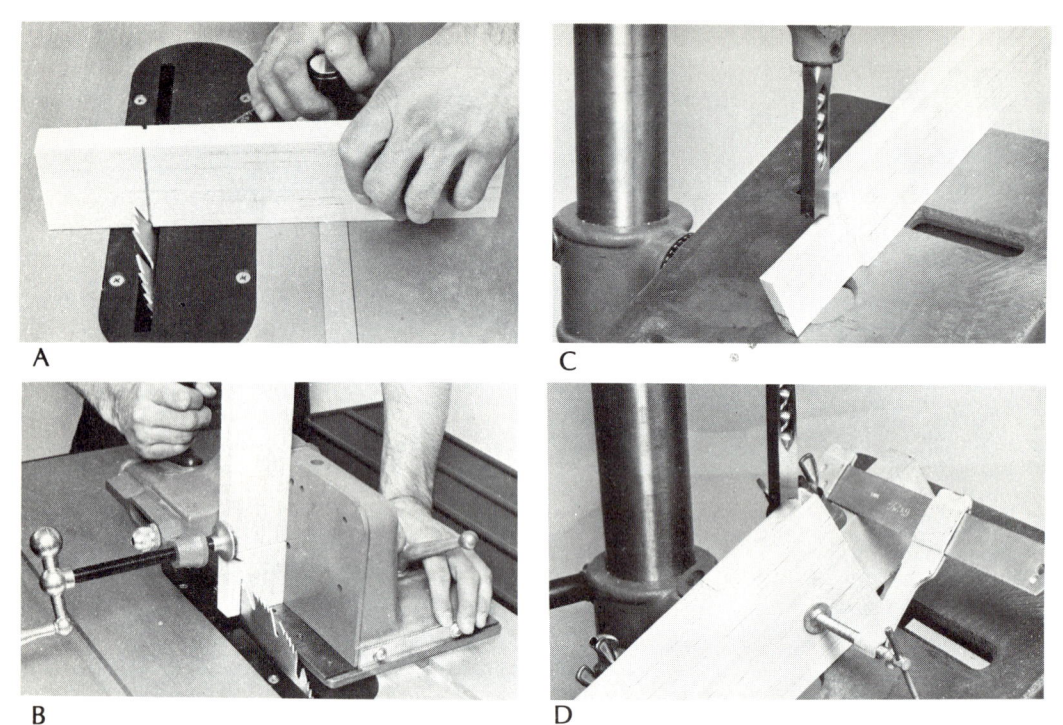

D

Fig. 4-9: Steps in cutting both the tenon and mortise for the concealed haunched mortise and tenon joint. (Note: Blade guard was removed for picture clarity; cut should be made with all saw safety devices in place.)

TUSK TENON

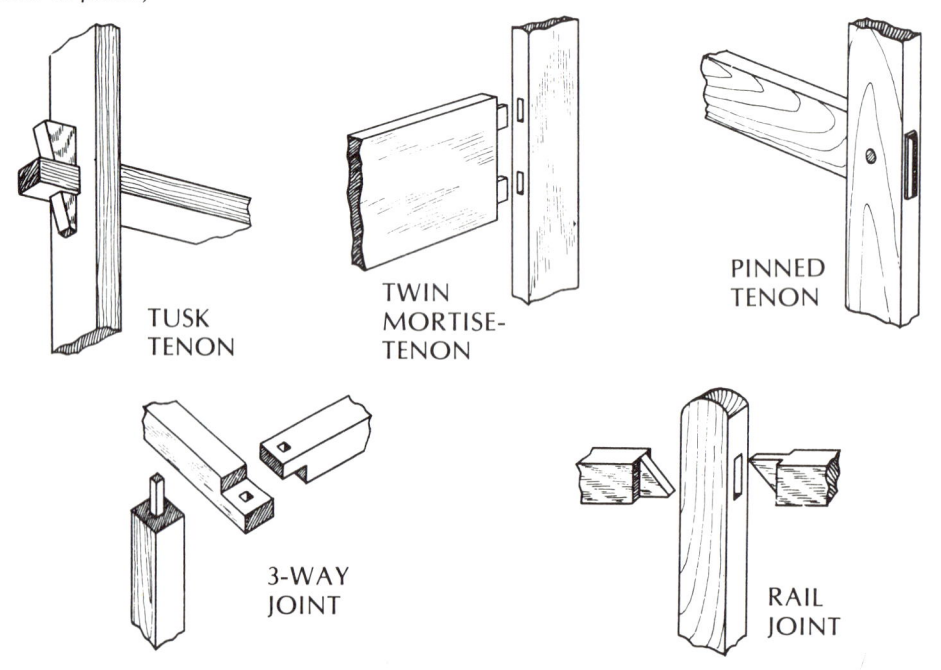

TWIN MORTISE-TENON

PINNED TENON

3-WAY JOINT

RAIL JOINT

Fig. 4-10: Other mortise-and-tenon joints.

76

tenon. The cut tenon is used as a guide for the drill press table tilt needed to cut the mortise (Fig. 4-9C). (If your table does not tilt, you can use the auxiliary tilting table described in Chapter 1.) The mortise being cut after two or three angle cuts is shown in Fig. 4-9D. The table is leveled to complete the main portion of the mortise. When using a radial type drill press, the head is tilted rather than the table.

Other frequently used mortise and tenon joints are shown in Fig. 4-10.

Side Mortise. The side mortise (Fig. 4-11) is often useful and is easily made by following the regular mortising sequence, except that a backing block is used against the work. Shallow side mortises or gains can be cut in the same manner, allowing the surplus width of the chisel to bite into the backing block (Fig. 4-12). A side mortise cut with a mortising chisel is always a little rough at the end, that is, at the bottom of the chisel cut. Where the joint shows and must be smooth, routing may be preferable.

Mortising Round Stock. To mortise rails into round legs or to install shelves on round posts by forming radial mortises, a V-jig such as the one shown in Fig. 4-13 may be used. The jig can also be used to cut mortises to attach corner-to-corner stretchers or rails when square legs are employed.

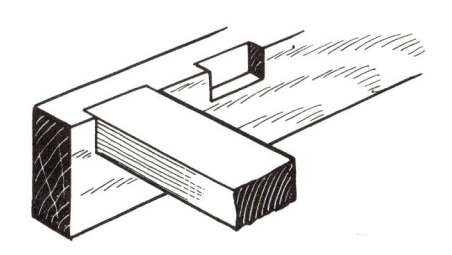

Fig. 4-11: A side mortise provides for flush rail insertion.

Fig. 4-12: A side mortise is easily cut with the use of a backing block.

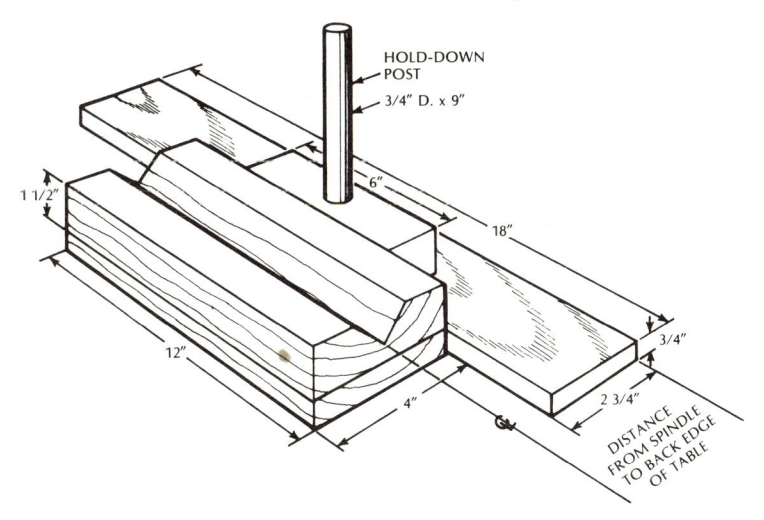

Fig. 4-13: Construction details of a V-jig for mortising.

When constructing the jig, make certain that the dimension from the center of the V to the back edge of the jig is equal to the distance from the center of the spindle to the back edge of the table. Also, be sure that the spindle and V have the same center line and that the chisel is square. Adhering to these dimensions will simplify the alignment process when employing the jig.

To use the jig, place the workpiece in the V and secure it with the standard mortiser hold-down (Fig. 4-14). Cut the stock in the normal procedure, making end cuts first, and then remove the material between.

Odd-shaped Work. Occasionally a mortise is required on odd-shaped work (Fig. 4-15). Sometimes such work makes it necessary to use blocks for support. An example is the leg shown in Fig. 4-16. Note

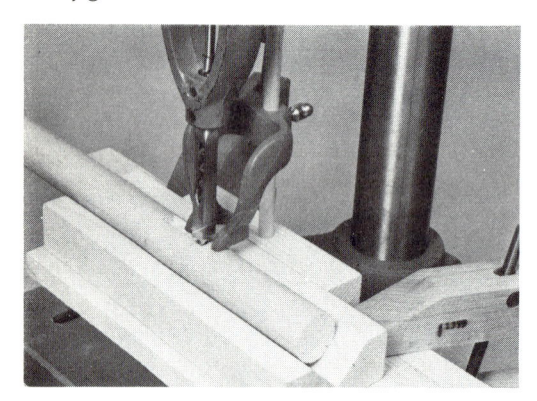

Fig. 4-14: V-jig in use.

Fig. 4-16: Odd-shaped work may require blocking and the use of a special fence.

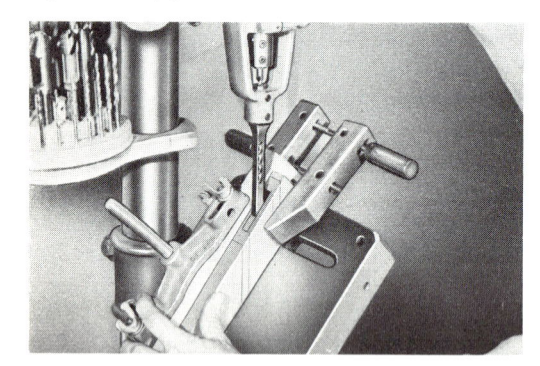

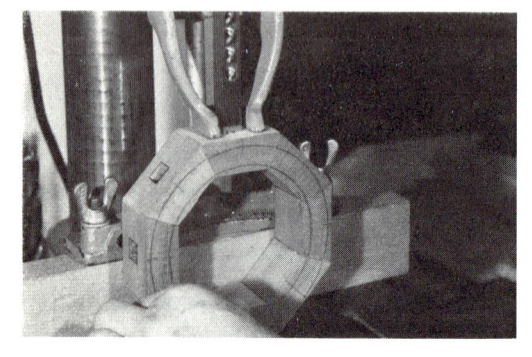

Fig. 4-15: If it can be properly held, it can be mortised.

Fig. 4-17: Square cutouts are easy to make with a mortiser.

the arrangement of the fence to provide a hold-down. On cutaway work like this and also on round legs, it is sometimes advisable to cut the mortises while the work is in the square (Fig. 4-17).

Mortising can also be employed to make various interior cutout designs on wood stock. In fact, the square cutout of the mortiser provides a wide variety of interesting patterns, including those of diamonds (Fig. 4-18). To make this latter shape, the mortising chisel must be turned in the chisel holder and then used in the normal fashion. After the diamond mortise cuts are made, strip-cutting on the table saw produces the thin strips that can be joined edge to edge to make a continuous pattern molding (Fig. 4-19).

BORING MORTISES

When regular mortising equipment is not available, very good mortises with round ends can be bored with a machine spur bit. The first hole is bored at the point where the left end of the mortise is to be cut. The stock is then moved a distance equal to 3/4 of the diameter of the bit, and another hole is bored. This is repeated until the full length of the required mortise has been bored, as shown in Fig. 4-20.

Remove the bit and replace it with a suitably-sized rotary rasp. Lower this to the bottom of the end hole and with the quill locked to maintain this position, the stock is pushed into the rasp the full length of the

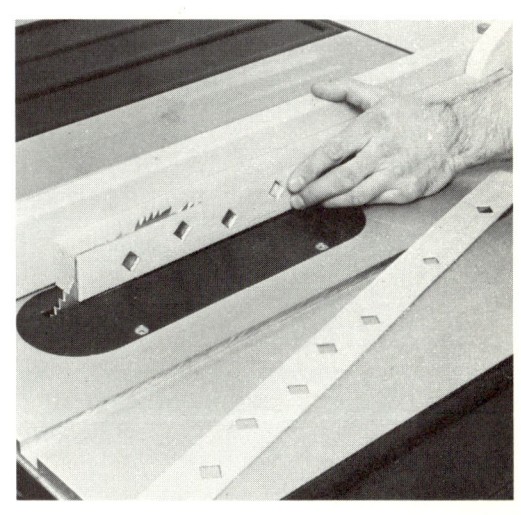

Fig. 4-19: Strips are cut on a table saw. (Note: Blade guard was removed for picture clarity; cut should be made with all saw safety devices in place.)

Fig. 4-18: The mortising chisel can be turned in the chuck to obtain diamond shapes.

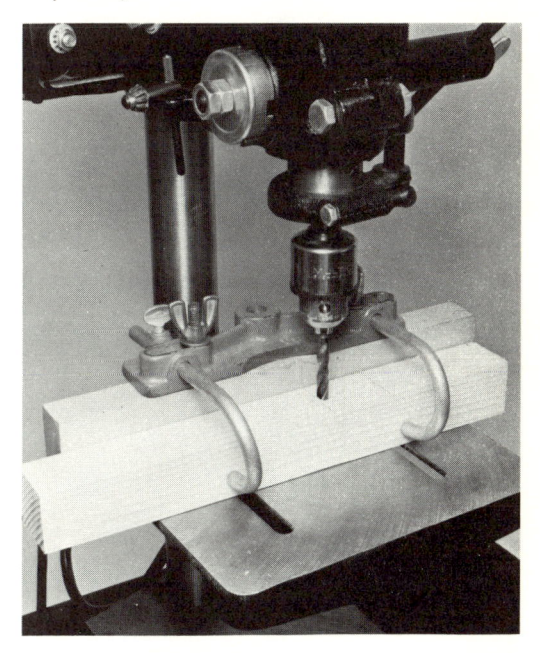

Fig. 4-20: When bored, the resulting mortise has round ends. The tenon must be rounded off to fit properly.

79

mortise (Fig. 4-21A). This cleans out the cut and gives a mortise similar to the one shown in Fig. 4-21B. Quite deep mortises can be made in this way. A spur bit must be used. Another method of boring is shown in Figs. 4-21C and D. In this case, holes are bored fairly close together (Fig. 4-21C), then other holes are drilled between, as shown by the dotted lines in Fig. 4-21D. The completed mortise is shown in Fig. 4-21E. The rasp is used as before to clean the mortise (Fig. 4-22).

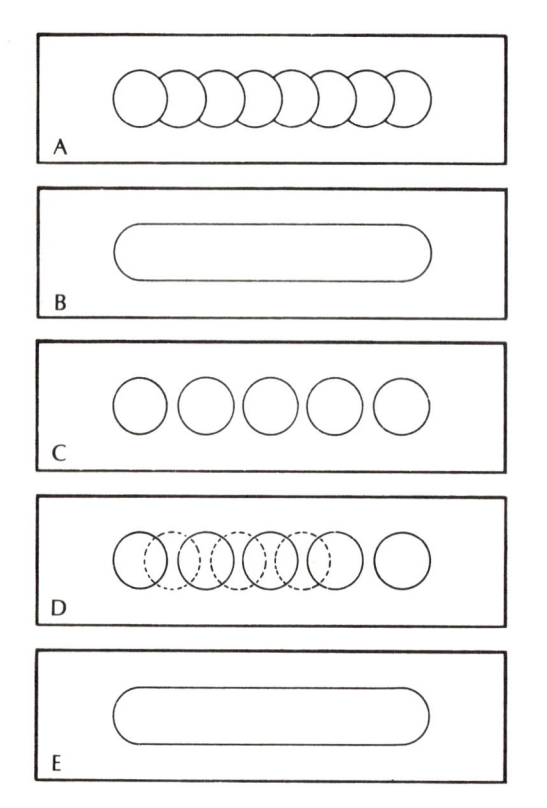

Fig. 4-21: Other methods of boring a mortise.

Fig. 4-22: Cleaning a mortise with a rotary rasp.

Chapter 5

DRILLING IN METAL

Drilling in metal is much the same as boring in wood, with certain differences. Since the material is much tougher and harder, speeds and feeds are more critical to the quality of the work and the life of the cutting tool. When drilling metal, it is essential to establish firm support for the work as close to the cutting area as possible. A drill turning in a piece of metal exerts a strong force which tends to rotate the work, the twisting action being most severe when the drill breaks through the underside. Naturally, a piece of work spinning on the end of a drill is not desirable—it can cut your hand, break the drill, and spoil the work. This hazard is avoided by clamping the work or otherwise mounting it to prevent twisting. Although it can be safe to hold the work by hand if the drill is small or if the work provides a firm hand grip several inches from the drilling point, it is always much safer to clamp a workpiece.

There are simple devices available that can either stop rotation or clamp the work securely for drilling. Some of the more popular methods are described here.

Stop Bolt. The simplest method of preventing the rotation of the workpiece is the stop bolt. Actually this "stop bolt" can be a machine bolt, a block of wood, the side of a clamp, a fence, or anything that provides a stop and prevents the work from twisting. One stop is sufficient for most work. Two stop bolts are sometimes used when drilling flat stock (Fig. 5-1), since they can be arranged to center the work crosswise under the drill. The drill column itself can be used as a stop. A most practical stop (Fig. 5-2) is the wrench or key for the geared chuck, since it is always available. A 5/16-inch hole will accept the key handle. The hole is drilled with the drill press itself by turning the table.

Fig. 5-1: A nut and stop bolt setup is a good way to counteract twist when drilling metal.

Fig. 5-2: A chuck wrench can be used as a stop.

Strap Clamps. The simple strap clamp (Fig. 5-3) is extensively used for holding work on all kinds of machine tools. The best type is U-shaped, because it allows the clamping bolt to take any position. A flat piece of steel with two or more holes along its length is often used. The best bolts for use with strap clamps are carriage bolts. Each bolt should have a large washer, big enough to span the ribs on the underside of the drill table, under the head.

Parallels. Holes can be drilled under two conditions. First, the work may be placed directly on the table, the drill passing through it and into the table opening. Second, the work may be such that it cannot be centered, in which case it is mounted on a wood block or on machined pieces of bar stock, called parallels, so that the drill will not strike the table after passing through the work. For average work, wood-base blocks serve quite well, but for more accurate drilling, carefully machined parallels should be used. When using parallels of any kind, the depth stop should be set so that the drill cannot run into the table at the completion of the drilling operation.

Standard Fixtures. Standard fixtures used in metal drilling are shown in Fig. 5-4. The rings at A are parallels cut from pipe stock. B shows the familiar type of bar parallel. C is a parallel which can be adjusted to different heights, often useful for supporting odd-shaped work. D is a plate clamp, which is very similar to the strap clamps already described except that it is fashioned from flat stock. E is a step block, the various steps being used in accordance with the height of the work. The step block is preferably of metal but good work can be done with a block cut from hardwood. Various kinds of V-blocks, used for centering round work, are shown in F, G, H, and I. The simplest type, F, is made from either wood or metal. G is another simple type made from two lengths of angle iron mounted on a wood base. Another homemade V-block is made by riveting two lengths of pipe to a base block, as in H. A factory-made style of V-block (Fig. I) is

Fig. 5-3: Manner of using stop bolts and strap clamps to hold work while drilling.

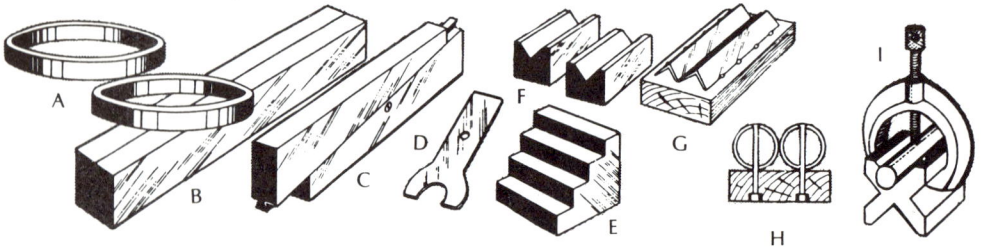

Fig. 5-4: Standard fixtures used for holding and clamping work.

fitted with a clamp to hold the work securely in place.

Guarding Against Spring. If a light piece of work is supported at points too far apart, it will spring under the pressure of the drill (Fig. 5-5). This may cause jamming or even damage to the drill when it breaks through. Always support the work sufficiently close

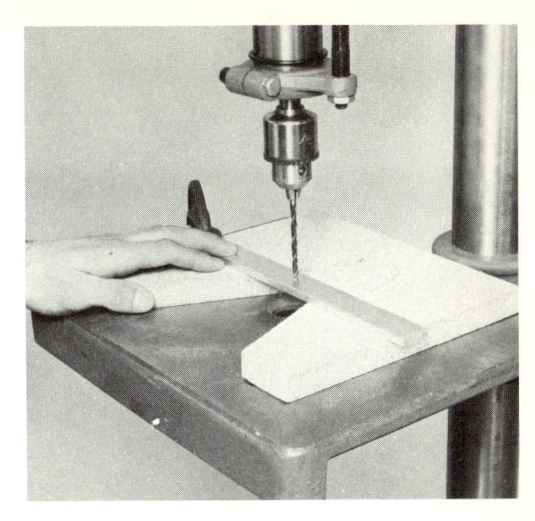

Fig. 5-7: A V-support is a good way to prevent spring.

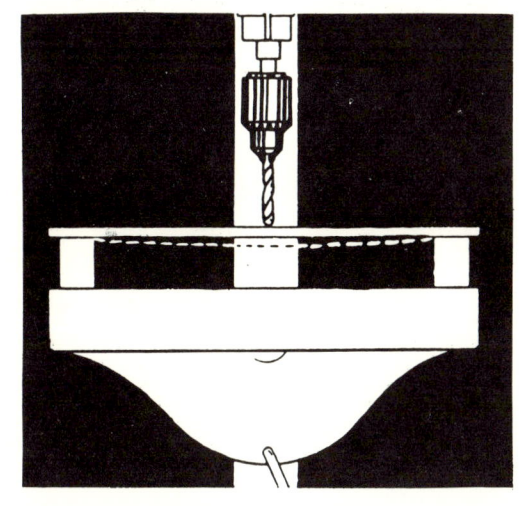

Fig. 5-5: The effect of spring.

to the hole being drilled so that it cannot spring. With this in mind, secure the workpiece with a C-clamp as shown in Fig. 5-6 and drill through the metal and into the wood. Stop drilling when wood chips appear. If you wish to avoid the wood chips, use a V-support as in Fig. 5-7.

Mounting Methods. There are various methods of mounting metal workpieces.

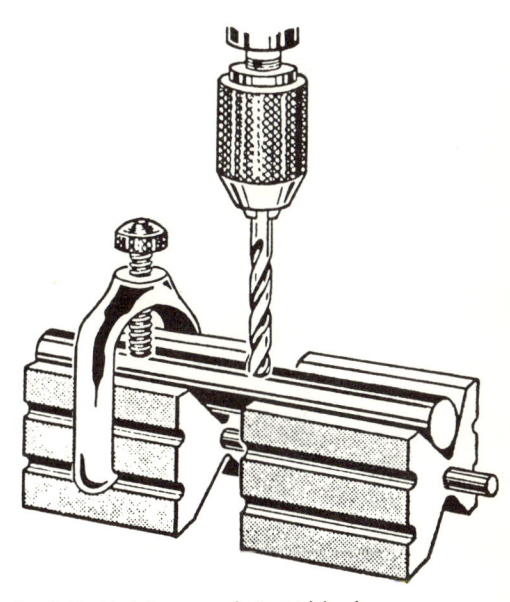

Fig. 5-8: Holding work in V-block.

The V-block arrangement illustrated in Fig. 5-8 is ideal for round work. The blocks are usually clamped to the table with C-clamps or held in position by stop bolts. An angle plate, which can be bolted or clamped to the drill press table, is useful for holding odd-shaped work (Fig. 5-9A). An application of adjustable parallels is shown in Fig. 5-9B. One method of setting up

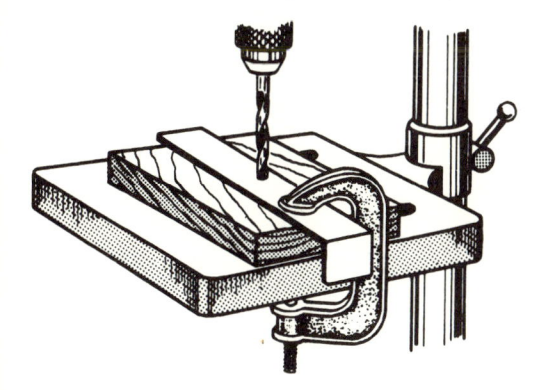

Fig. 5-6: One way of counteracting spring.

83

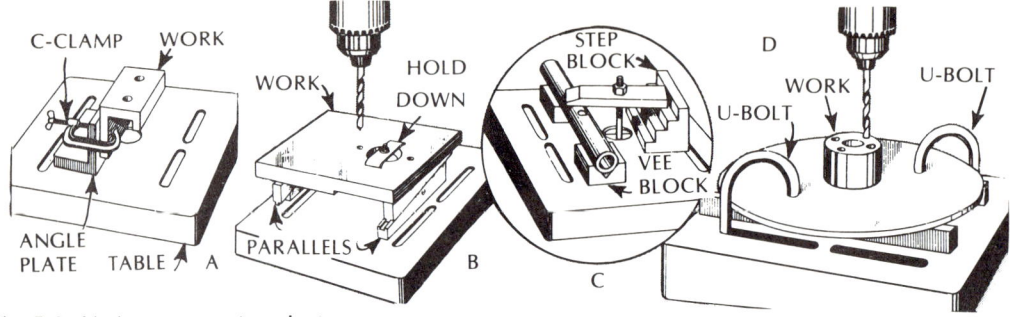

Fig. 5-9: Various mounting devices.

round work for drilling is shown in Fig. 5-9C. U-bolts or hook-bolts (Fig. 5-9D) are often useful. Numerous other methods are used in mounting work for drilling, most of the setups being of a nature more suitable for production drilling than homeshop work.

Use of the Vise. An adjustable vise offers the fastest and best method of mounting most drill press jobs (Fig. 5-10). The V-grooves in the jaws will hold round work in either a vertical or horizontal position (Figs. 5-11 and 5-12). For heavy work or repeat drilling, the vise should be bolted to the drill press table, (Fig. 5-13). While it is sometimes safe to hold the vise by hand, it is always safer to use a stop bolt or block to prevent rotation.

When mounting flat or square work in the vise, the usual method is to place the

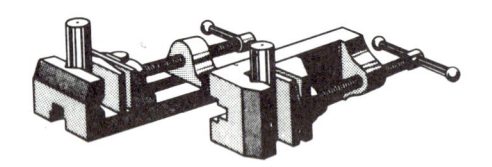

Fig. 5-11: Holding work to drill holes in the end of round stock.

Fig. 5-12: The V-grooves in the vise jaws hold round work easily.

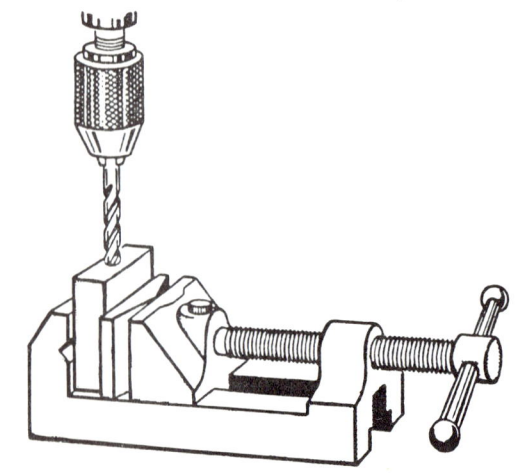

Fig. 5-10: Holding work with a drill press vise.

Fig. 5-13: A vise can be bolted to the drill press table.

work at the bottom of the jaws and in such a position that the drill will clear in the center channel of the vise. Some workers prefer mounting the work flush with the top of the vise jaws, without support. In a third method of mounting, a wood block or parallels are placed under the work (Fig. 5-14). This is probably the best method since it supports the work, allows a clear view, and also provides a base for the breakthrough.

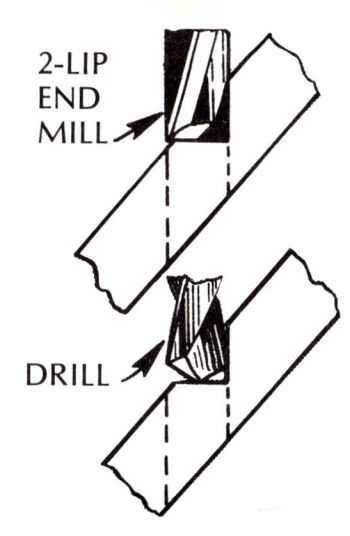

Fig. 5-15: An end mill cuts into a slanting surface without walking, providing a flat for drilling.

Fig. 5-14: A wood block gives support to the drilling.

Angular Drilling. Many drill vises have an angle adjustment which permits any degree of tilt needed for angular drilling. When the hole must be entered on a slanting surface, the best work method is to use an end mill to establish a flat surface and then follow with the drill (Fig. 5-15). When work of this kind must be done with an ordinary twist drill, one method is to start the hole with the drill square to the workpiece.

Clamping Jig. As versatile as the drill vise is, it is less suitable in a large number of drilling operations, including holding odd shapes, thin metal, and wide work. A popular device for work of this kind is a hand-held clamping jig (Fig. 5-16). It can be used for light drilling. The construction is shown in Fig. 5-17. Initial clamping pressure is obtained by turning the wing nut, after which a strong leverage action is attained by turning the clamp screw handle. The 1/4-inch plywood pad over which drilling is done will eventually become riddled with holes but is easily replaced.

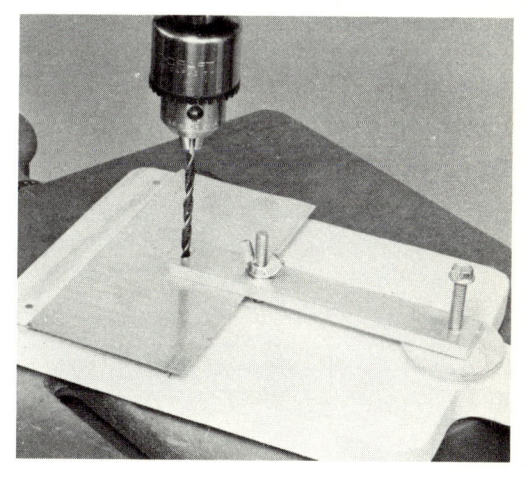

Fig. 5-16: The hand-held clamping jig is a very popular method work.

85

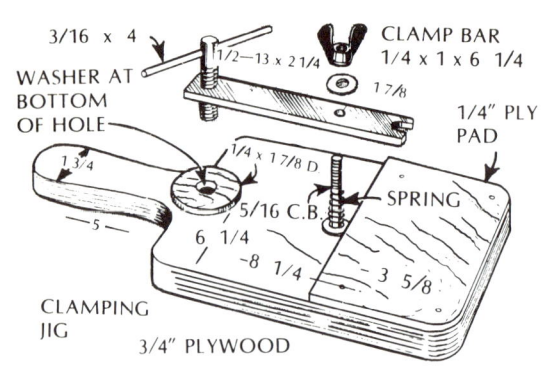

Fig. 5-17: Construction details of clamping jig.

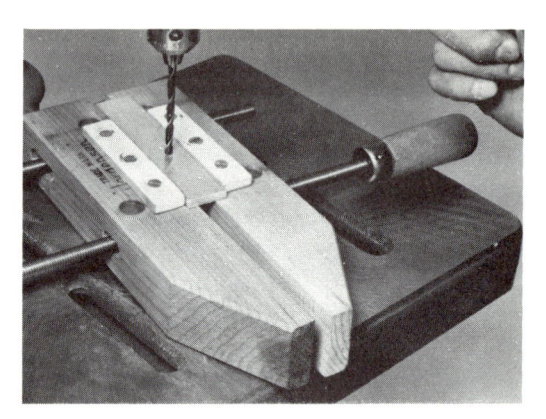

Fig. 5-19: Thin metal material can be supported by handscrew jaws.

DRILLING OPERATION

Twist drills are used for metal drilling operations. As mentioned in Chapter 2, these drills are available in number, letter, fractional, or metric sizes (see page 12). Since most homeshops do not have all types, the chart of decimal equivalents can be used to select a size that comes close to what is needed.

DECIMAL EQUIVALENTS

1/64 = .015625	3/4 = .750
1/32 = .03125	49/64 = .765625
3/64 = .046875	25/32 = .78125
	51/64 = .796875
1/16 = .0625	13/16 = .8125
5/64 = .078125	53/64 = .828125
3/32 = .09375	27/32 = .84375
7/64 = .109375	55/64 = .859375
1/8 = .125	7/8 = .875
9/64 = .140625	57/64 = .890625
5/32 = .15625	29/32 = .90625
11/64 = .171875	59/64 = .921875
3/16 = .1875	15/16 = .9375
13/64 = .203125	61/64 = .953125
7/32 = .21875	31/32 = .96875
15/64 = .234375	63/64 = .984375
1/4 = .250	1/2 = .500
17/64 = .265625	33/64 = .515625
9/32 = .28125	17/32 = .53125
19/64 = .296875	35/64 = .546875
5/16 = .3125	9/16 = .5626
21/64 = .328125	37/64 = .578125
11/32 = .34375	19/32 = .59375
23/64 = .359375	39/64 = .609375
3/8 = .375	5/8 = .625
25/64 = .390625	41/64 = .640625
13/32 = .40625	21/32 = .65625
27/64 = .421875	43/64 = .671875
7/16 = .4375	11/16 = .6875
29/64 = .453125	45/64 = .703125
15/32 = .46875	23/32 = .71875
31/64 = .484375	47/64 = .734374

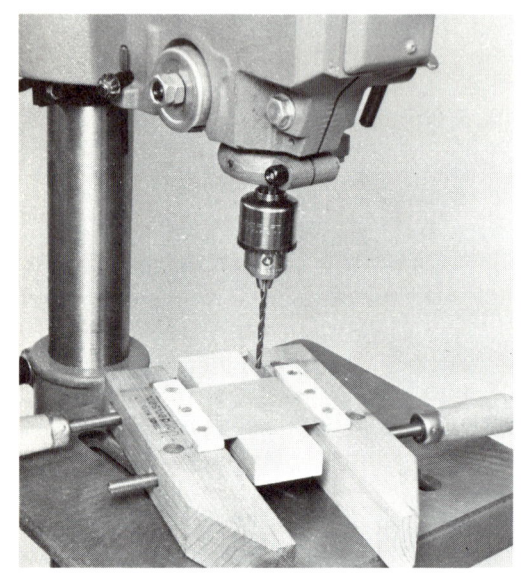

Fig. 5-18: A handscrew clamp makes a most useful mounting device; the long handles provide plenty of leverage to resist the twisting action of the drill.

Mounting with a Handscrew. Another useful device for mounting small work or sheet metal is a handscrew clamp fitted with two wood or metal strips fastened to the regular jaws. The method of using it is shown in Fig. 5-18. Sheet metal stock can be supported directly on a wood block placed over the screws (Fig. 5-19). Many other uses will be found for the handscrew in drilling when the torque action is not severe. Most hand-held clamping devices are satisfactory with drills up to about 1/4-inch diameter.

Drilling Procedure. Several methods are used in the actual drilling of the hole. These methods are essentially a matter of getting the drill centered at the required location. One system used for exacting work is done with an accurate metal pointer or scriber chucked in the drill chuck, as shown in Fig. 5-20. If the pointer or scriber is carefully centered over the required hole positon, the drill which takes its place will also be centered. Work of this kind is often done without the use of a punch mark, although if the hole is large it is more accurate to start it with a center drill or a smaller drill. When two or more tools must be chucked in a setup of this kind without moving the table, the setup should be checked by chucking the longest tool in the drill chuck.

Fig. 5-21: The drill engaged in a conical recess of partly-drilled hole holds the workpiece in place while the clamp is applied.

Fig. 5-20: Pointer method of centering work.

When the work is to be clamped by any method, the hole should be started by holding the work by hand. Then, with the drill seated in the conical recess, clamping pressure is applied (Fig. 5-21), being careful not to disturb the position of the work. The same procedure is followed if a pointer is used to locate the work position.

The most common method of centering the work is done freehand by holding the work lightly to allow the drill to center on the punch mark. The punch mark can guide the drill only if it is larger than the web of the drill (Fig. 5-22). This does not necessarily mean that a big punch mark is needed, since the web of a 1/4-inch drill is no more than 1/32 inch long. When a large hole is being drilled, it is usually best to spot and drill the hole with a smaller drill rather than trying to make a punch mark big enough to take the web of the larger drill. This whole procedure is shown in Fig. 5-23. An alternate method of locating the center line on round work is shown in Fig. 5-24. The drill is brought down lightly on the work and clamped. Then, the work is pushed under the drill, which will make

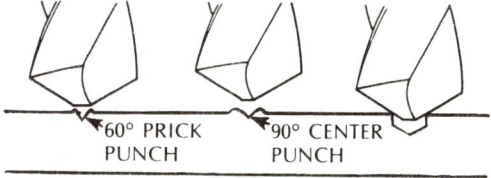

60° PRICK PUNCH   90° CENTER PUNCH

PUNCH MARK TOO SMALL   PUNCH MARK RIGHT SIZE   SMALL DRILL USED TO START

Fig. 5-22: How to determine proper punch mark.

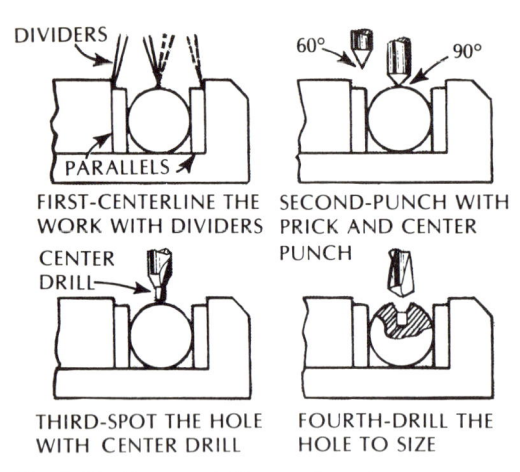

FIRST—CENTERLINE THE WORK WITH DIVIDERS
SECOND—PUNCH WITH PRICK AND CENTER PUNCH
THIRD—SPOT THE HOLE WITH CENTER DRILL
FOURTH—DRILL THE HOLE TO SIZE

Fig. 5-23: How to drill round work.

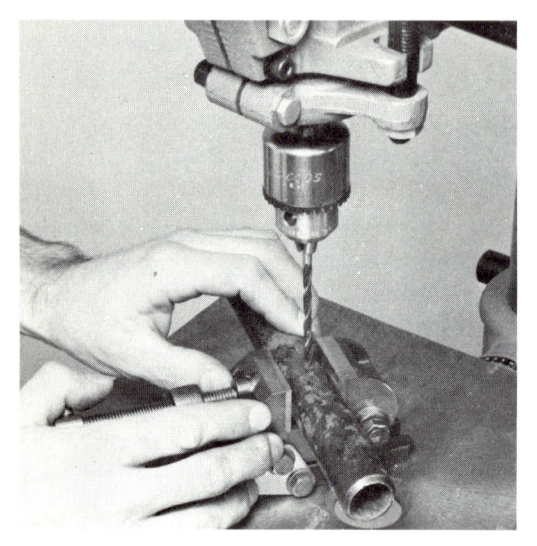

Fig. 5-24: Revolving drill just touching the work marks the center for punching.

a short centering mark on the top of the workpiece.

Despite care in laying out and clamping, sometimes the hole may be off-center after cutting a few revolutions. How to "draw" a hole that has run off is shown in Fig. 5-25. A small, shallow groove is cut with the use of a small, round-nose chisel, the position of the groove being on the side toward which it is desired to draw the hole. When the drill is started again, it should drift to the correct position. If not, the chisel must be used again. Drawing must be done before the drill starts to cut its full diameter. You can use a square-nose chisel if you like, holding it with the flat edge against the drilled cone. However, be sure to cut out enough to drift the bit back to the center when the operation is resumed.

Some metals can be drilled without coolant, but kerosene or cutting oil should be applied to ferrous metals (especially the harder steels), to prevent overheating of the drill point.

Place a drop or so of coolant on the spot before starting. Then add coolant as needed. Chip clearing is also important particularly when drilling deep holes. Raise the drill frequently, and do not allow the flutes to clog with chips.

Proper operating speeds will vary with the size of the drill and the material being worked. A general guide to operating speeds can be found on page 40, while the table on the next page is more exact. If the speed suggested is not available in either chart, use the closest *lower* speed possible. Also watch the rate of feed. If the feed pressure is too heavy, the chips will tend to

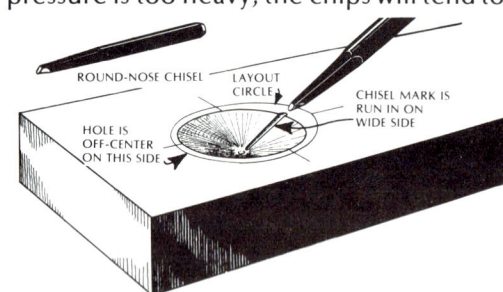

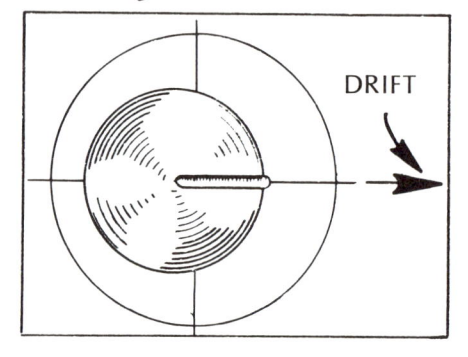

Fig. 5-25: How to "draw" a hole which has run off-center.

DRILL SPEEDS IN R.P.M.
For Various Materials

Diameter of Drill	Soft Metals 300 F.P.M.	Plastics and Hard Rubber 200 F.P.M.	Annealed Cast Iron 140 F.P.M.	Mild Steel 100 F.P.M.	Malleable Iron 90 F.P.M.	Hard Cast Iron 80 F.P.M.	Tool or Hard Steel 60 F.P.M.	Alloy Steel Cast Steel 40 F.P.M.
1/16 (No. 53 to 80)	18320	12217	8554	6111	5500	4889	3667	2445
3/32 (No. 42 to 52)	12212	8142	5702	4071	3666	3258	2442	1649
1/8 (No. 31 to 41)	9160	6112	4278	3056	2750	2445	1833	1222
5/32 (No. 23 to 30)	7328	4888	3420	2444	2198	1954	1465	977
3/16 (No. 13 to 22)	6106	4075	2852	2037	1833	1630	1222	815
7/32 (No. 1 to 12)	5234	3490	2444	1745	1575	1396	1047	698
1/4 (A to E)	4575	3055	2139	1527	1375	1222	917	611
9/32 (G to K)	4071	2712	1900	1356	1222	1084	814	542
5/16 (L, M, N)	3660	2445	1711	1222	1100	978	733	489
11/32 (O to R)	3330	2220	1554	1110	1000	888	666	444
3/8 (S, T, U)	3050	2037	1426	1018	917	815	611	407
13/32 (V to Z)	2818	1878	1316	939	846	752	563	376
7/16	2614	1746	1222	873	786	698	524	349
15/32	2442	1628	1140	814	732	652	488	326
1/2	2287	1528	1070	764	688	611	458	306
9/16	2035	1357	950	678	611	543	407	271
5/8	1830	1222	856	611	550	489	367	244
11/16	1665	1110	777	555	500	444	333	222
3/4	1525	1018	713	509	458	407	306	204

Figures are for High-Speed Drills. The speed of Carbon Drills should be reduced one-half. Use drill speed nearest to figure given.

build up rapidly and will clog the drill. When too light a feed pressure is used, the drill will not cut. Keep the feed pressure uniform until the drill is about to break through; then reduce the pressure slightly for the breakthrough, because this is the critical point where it is most likely for the drill to grab or break.

Drilling Sheet Metal. The nature of sheet metal demands that the work be supported on a wood base. To prevent the material from spinning, it is necessary to provide a stop against rotation, such as the fence in Fig. 5-26. Where many holes are being drilled, it is also worthwhile to provide some kind of hold-down to prevent the natural tendency of the thin metal to climb the drill after the hole has been made.

Much of the work in sheet metal is concerned with assemblies, such as boxes, cones, and so on. To provide a support inside the work, it is necessary to use some kind of extension bar or stake clamped to the drill table or held in a vise clamped to the drill table. A typical arrangement is shown in Fig. 5-27. One end of the bar is

METAL-DRILLING PROBLEMS AND SOLUTIONS

The following chart lists some common problems that are sometimes encountered in drilling metal and their possible solutions:

Drilling Problem	Cause	What To Do
Drill overheats and turns blue at the corners.	Cutting speed too high.	Reduce spindle speed
Drill breaks or chips at cutting edges.	Excessive clearance.	Regrind drill; use less clearance.
Drill slits up the web.	Feed rate too high; insufficient clearance.	Reduce feed rate; regrind drill; use more clearance.
Hole drilled oversize (it should be understood that a drill normally cuts slightly oversize).	Unequal lip lengths; tool misaligned in drill holder.	Regrind drill; use a drill-point gauge to check for correct angle and lip length; realign tool or change drill holder.
Hole drilled undersize (common with abrasive materials such as phenolic resins, hard rubber, aluminum castings, and cast-iron scale).	Margin has been worn down.	Replace drill.
Bell-mouth hole.	Drill not held properly in machine; work not held securely.	Check and adjust drill clamping; reclamp work securely.

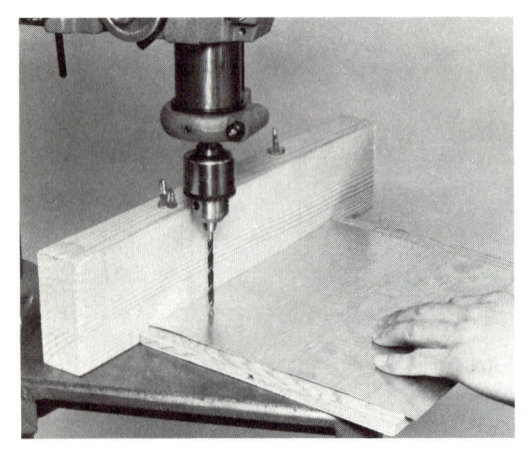

Fig. 5-26: A fence assures accuracy and also guards against work rotation when drilling sheet metal.

Fig. 5-27: An extension arm is useful for assembled work.

fitted with a small wood drill pad, while the other end has a smaller support consisting of a carriage bolt set in a hole drilled partway through the arm. This support is being used in the photo (Fig. 5-27).

The major problem when drilling sheet metal is the formation of the burrs on the underside of the work. For occasional work this is simply cleaned up with a countersink, but, for long runs, it is often worthwhile to use specially ground drill points which eliminate or at least reduce burring. The simplest of these is the blunt point shown in Fig. 5-28. It cuts a clean hole but tends to walk unless the point is thinned down, as shown, to center readily on the punch mark. The spur style cuts a clean hole. It is actually a small disk cutter and forms the hole by cutting out a disk rather than by drilling. The only fault with this style is that the disk often sticks to the end of the drill. The flat style is simply an extreme of the blunt point idea; you are cutting the full diameter right from the start, with none of the wedging action of a cone point. This style needs a brad point for centering, which makes it somewhat difficult to grind. The various points shown are suitable for drills under a 3/4-inch diameter. Large holes in sheet metal are more readily cut with the use of a hole saw (Fig. 5-29). The hole saw is described on page 6.

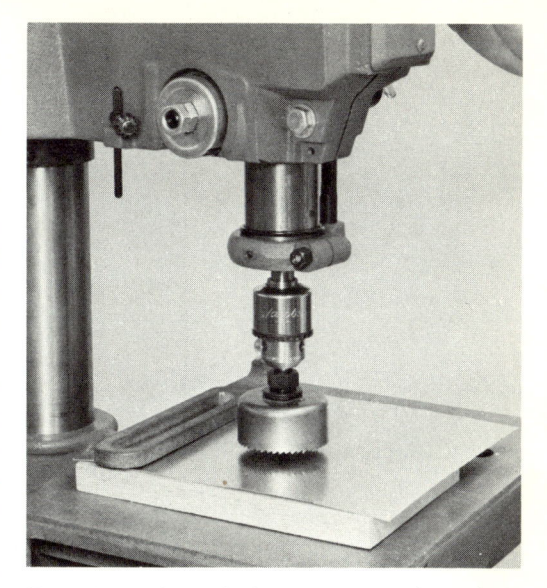

Fig. 5-29: Using a hole saw to cut a large hole in sheet metal.

Large Holes. Large holes can be made on the drill press, more accurately and easily, if a small hole is drilled first. The lead hole should not be smaller than the web thickness of the larger drill. Where accurate centering is critical, it is best to start the hole with a large drill, drawing as needed to center. The countersink made with the large drill will center the pilot drill. After the small hole is through, the larger drill is again used and the hole drilled full-size.

Mounting Work Vertically. Where end drilling is to be done, the work does not usually possess sufficient rigidity to stand with any degree of accuracy. The common solution to this problem is the drill press vise (see page 18), and good work can be done at extensions of several inches. An auxiliary tilting table can also be employed. If you have a tilting table, use it in a vertical position. A length of angle iron clamped to the lower part of the table supplies a convenient base on which to rest the work (Fig. 5-30). Heavy work should also be clamped, as shown, although light drilling can be done without clamping because the drill table itself supplies a stop against

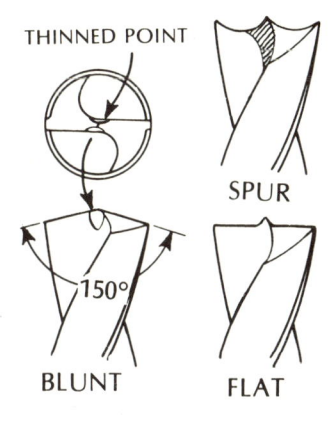

Fig. 5-28: Points designed to reduce burring.

Fig. 5-30: A drill table tilted to a vertical position provides a good support for end drilling.

rotation. Some of the setups described for drilling in wood can also be used, page 70.

As when boring wood, the radial drill press is most handy for horizontal work and the angle drilling of metals that can be drilled within the capacities of the machine. When working on metal, the head should be as close as possible to the column to increase the rigidity of the machine (Fig. 5-31).

COUNTERSINKING AND COUNTERBORORING

Various types of machine screws are frequently set so that their heads come flush or below the surface of the work (Fig. 5-32). A satisfactory countersink for

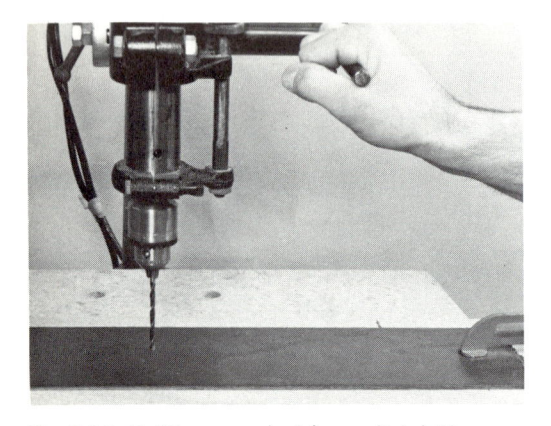

Fig. 5-31: Drilling metal with a radial drill press.

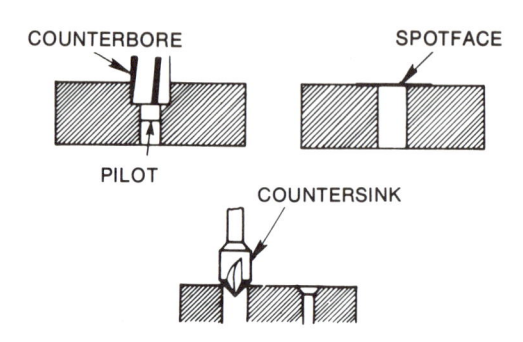

Fig. 5-32: Methods of setting machine screws flush or below th surface of the work.

machine screws can be made by grinding an ordinary twist drill to the required point angle. A wide variety of special countersinks can be purchased for this work, two typical styles being shown in Fig. 5-33. These bits are frequently used to remove the sharp edge from a hole prior to tapping or reaming.

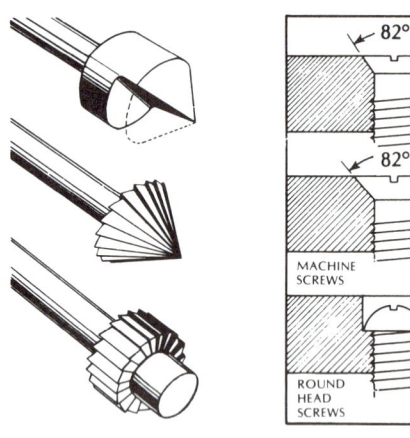

Fig. 5-33: Two countersinks and a counterbore that are used for machine screws.

When countersinking a flat-head machine screw, select a countersink bit that has a diameter at least as large as the diameter of the top of the screwhead. The spindle speed should be set for that diameter of bit. When sinking the screwhead exactly flush, adjust the depth stop so the countersink bit will enter the work to the point at which its diameter is exactly the same as the screwhead top.

TAP DRILL SIZES AND PERCENTAGES OF THREAD

	NATIONAL COARSE THREAD				NATIONAL FINE THREAD		
Tap	90%	75%	50%	Tap	90%	75%	50%
2-56	No. 51-83%	No. 50-70%	No. 49-56%	2-64	No. 50-80%	No. 49-65%	5/64-40% No. 48-50%
3-48	No. 48-85%	No. 47-78% 5/64-77%	No. 44-50%	3-56	5/64-91%	No. 45-74%	No. 44-56%
4-40	No. 44-81%	No. 43-72%	No. 41-50% 3/32-56%	4-48	No. 43-85%	No. 42-70% 3/32-70%	No. 40-52%
5-40	No. 41-90% 3/32-97%	No. 38-75%	7/64-50%	5-44	No. 40-90%	No. 37-70%	No. 35-50% 7/64-53%
6-32	No. 37-83%	7/64-71% No. 36-78%	No. 32-53%	6-40	7/64-91%	No. 33-78%	1/8-41% No. 31-59%
8-32	No. 30-88% 1/8-90%	No. 29-68%	No. 27-49% 9/64-56%	8-36	No. 30-100%	9/64-64% No. 29-78%	No. 27-56%
10-24	No. 27-85% 9/64-91%	No. 25-76% 5/32-63%	No. 20-54%	10-32	5/32-83% No. 23-88%	No. 21-76%	11/64-44% No. 18-51%
12-24	No. 18-87%	No. 16-72% 11/64-81%	No. 12-50% 3/16-54%	12-28	No. 16-85%	3/16-61% No. 14-74%	No. 11-54%
1/4-20	No. 10-88% 3/16-97%	No. 6-71% 13/64-72%	7/32-47% No. 3-57%	1/4-28	No. 4-89%	7/32-67% No. 3-80%	No. 2-63% No. 1-48%
5/16-18	1/4-86%	17/64-65% F-76%	9/32-43% 1-55%	5/16-24	17/64-87%	1-74%	9/32-57%
3/8-16	N-90%	5/16-78%	Q-53% 21/64-58%	3/8-24	21/64-87%	Q-80%	S-50% 11/32-57%
7/16-14	23/64-84% T-85%	3/8-66% U-74%	25/64-50%	7/16-20	3/8-95%	25/64-72% W-78%	13/32-48% Y-51%
1/2-13	Z-87%	7/16-63% 27/64-79%	29/64-47%	1/2-20	7/16-97%	29/64-72%	15/32-48%
9/16-12	15/32-87%	31/64-72%	1/2-57%	9/16-18	1/2-86%	33/64-64%	17/32-43%
5/8-11	33/64-92%	35/64-67% 17/32-80%	9/16-53%	5/8-18	9/16-87%	37/64-65%	19/32-44%
3/4-10	41/64-85%	21/32-72%	11/16-48%	3/4-16	43/64-96%	11/16-78%	45/64-58%
7/8-9	3/4-87%	49/64-77%	13/16-44% 51/64-54%	7/8-14	51/64-84%	13/16-68%	53/64-51%
1-8	55/64-87%	7/8-77%	59/64-48% 29/32-58%	1-14	59/64-84%	15/16-68%	61/64-51%

A countersink plate saves time if much work of this nature is to be done. This is a flat metal plate, in which all of the standard countersinks used in average work, have been drilled. When a particular machine screw is to be fitted, it is first fitted into the correctly sized opening in the plate (Fig. 5-34). The plate is then placed on top of the work (Fig. 5-35) and the countersink brought down to seat in the selected hole. The depth stop is set at this position. If the plate is now placed below the work (Fig. 5-36), the countersink will cut to the proper depth.

The counterboring of machine screws and bolts in the home workshop can be done with a special counterbore bit or with

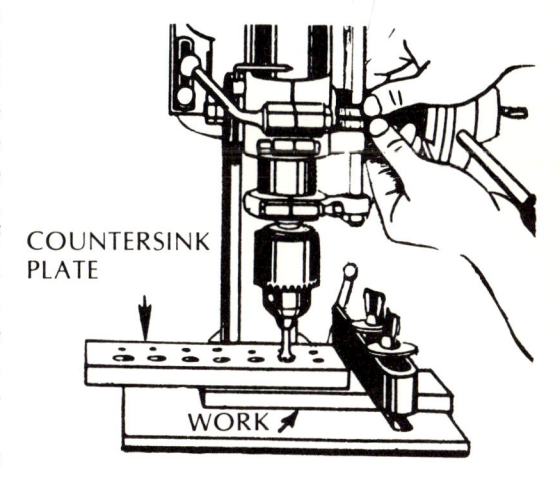

Fig. 5-34: A countersink plate is useful where much work of this kind is to be done.

93

Fig. 5-35: Matching a screw to its countersink.

Fig. 5-36: Countersinking with the countersink plate below the workpiece.

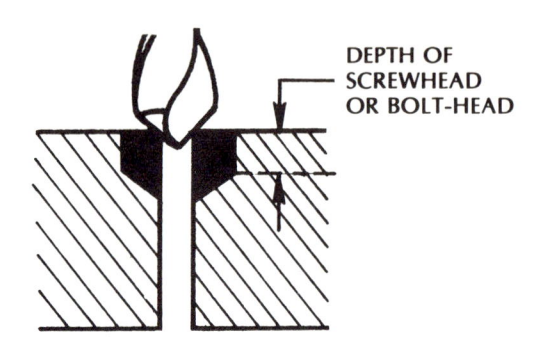

Fig. 5-37: Using a twist drill to make a counterbore.

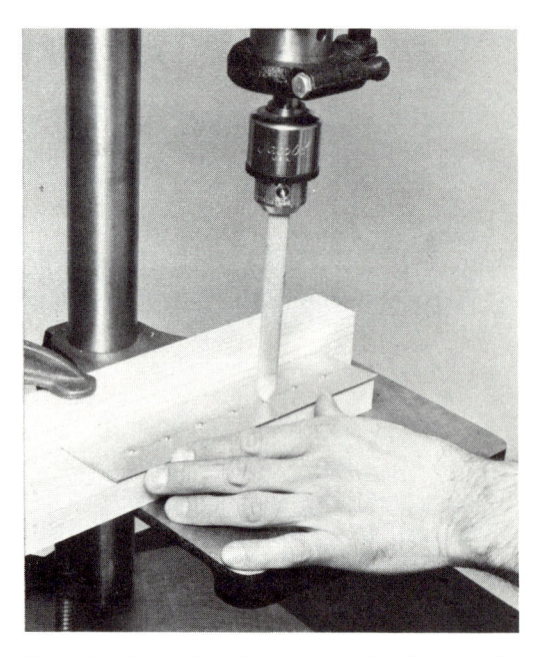

Fig. 5-38: Screwhead seats can be formed by "dimpling."

an ordinary twist drill. In either case, the bit or drill should be slightly larger than the screwhead or the bolt-head. When using a drill, it is best to have the recess shoulders slant down toward the center of the hole (Fig. 5-37), rather than attempting to make it flat. To recess the bolt or screwhead fully, adjust the depth stop so that the outer ends of the bit lips will penetrate to a depth at least equal to the screwhead or bolt-head height.

Spot-facing is similar to counterboring. It is done to produce a flat surface in a local area or a spot on which to seat a washer or bolt head. The same bits can be used for both counterboring and spot-facing.

When working with thin sheet metal, screwheads can be seated by "dimpling" the surface with a hardwood dowel on which the end is tapered (Fig. 5-38). Employ a slow speed when performing this operation and use wax as a lubricant.

REAMING AND LAPPING

Reaming and lapping are performed with special tools such as those shown in

Fig. 5-39. When a hole must be of precise diameter, it is first drilled undersized (approximately 1/64 inch smaller). Then a reamer, operated at slightly less than normal drilling speed, is employed to bring it to the desired diameter.

If a polished finish is required, the final enlargement is performed with a lap tool and the proper abrasive for the material being lapped. The lap must be the same diameter as the drilled or reamed hole. The tool is coated with the abrasive and lubricant, and operated at the *lowest* possible speed. Run the lap in and out of the hole with plenty of additional lubrication, until the desired final hole size and degree of polish are accomplished.

TAP DRILL TOO SMALL

TAP DRILL TOO BIG

TAP DRILL CORRECT

DRILLING FOR TAPPED HOLES

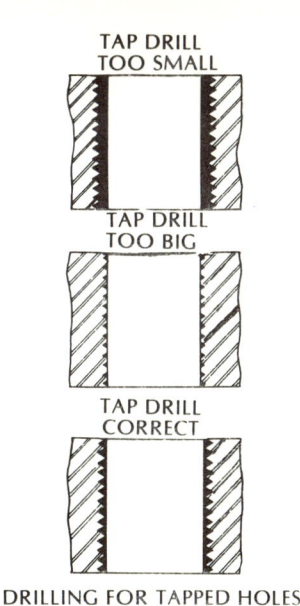

Fig. 5-40: The proper drill for a tapped hole should allow for 50 to 90 percent of a full thread.

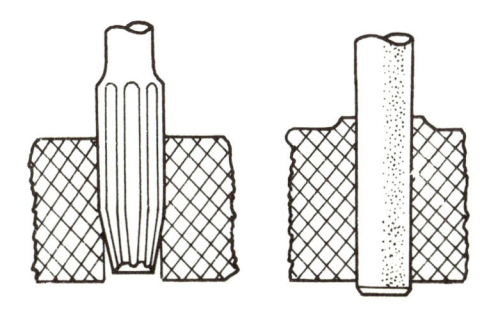

Fig. 5-39: The reamer (left) is for cutting, while the lap (which is a round metal rod) is for polishing a hole.

DRILLING FOR TAPPED HOLES

One of the major problems encountered when tapping holes is not keeping the tap square to the work. By using a drill press to do the job, you are assured of squareness. *The tapping process is not done under power, except when the proper equipment is available.*

The drill used for making a hole preparatory to tapping must be of certain definite size if good work is to be done. Consult the table on page 93 and use the drill specified. Figure 5-40 shows the effect of various sizes of tap holes. In the first example, the drilled hole is too small. The black area shows the amount of metal

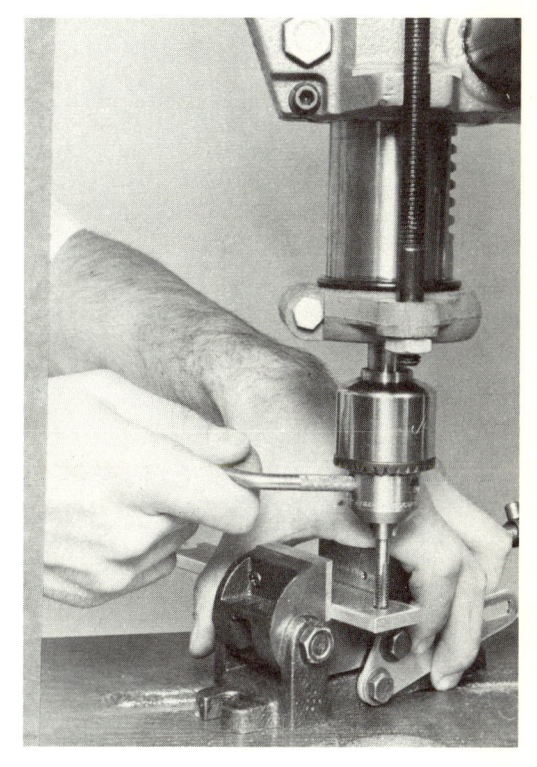

Fig. 5-41: The tapping operation is done by hand—never under power, not in the home workshop anyway.

95

which must be removed by the tap. Removal of the excess metal means a rough thread, premature dulling of the tap, and possible tap breakage. The center example is just the opposite. The thread here is only partly cut greatly weakening the holding power of the screw. The third example shows the right condition, a hole which will allow the tap to cut 70 to 75 percent of a full thread.

Starting the Tap. The tap can be started by mounting it in the drill press. To turn the chuck by hand, use a short length of metal rod or a suitable bolt in the chuck holes normally used by the chuck key (Fig. 5-41). While turning the chuck, apply a very light, even feed pressure. Some operators like to use a weight on the feed lever (Fig. 5-42) to give this even pressure. Always use plenty of threading or cutting oil on the tap when tapping steel. Cast iron is worked dry. A cleaner thread will be obtained if the tap is reversed to break the chip and remove the waste. For right-hand threads, turn the chuck counter-clockwise about one-fourth turn for every half turn clockwise. When withdrawing the tap from the tapped hole, be sure to keep some pressure on the spindle as the chuck is turned to clear the tap.

Tap Styles. Figure 5-43 shows the common taper tap, which is smaller at the end to permit easy starting. This tap must go completely through the work to cut a full thread. Blind holes are tapped with a plug tap. If the threads are to go completely to the bottom, the hole must be finish-tapped with a bottoming tap (Fig. 5-44).

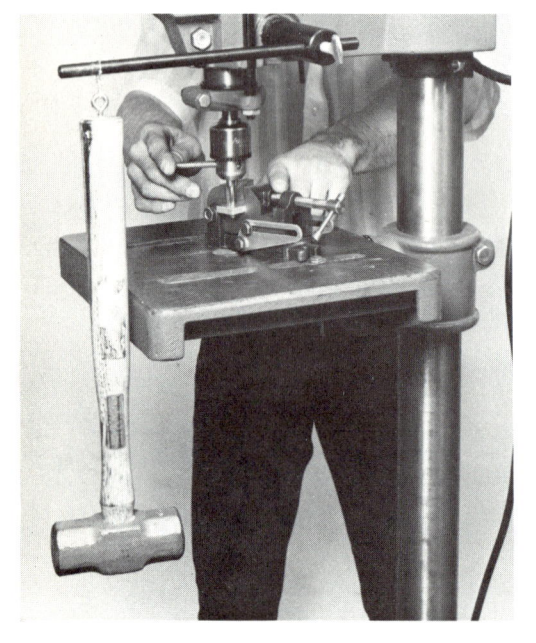

Fig. 5-42: Weight on the feed lever is useful when tapping.

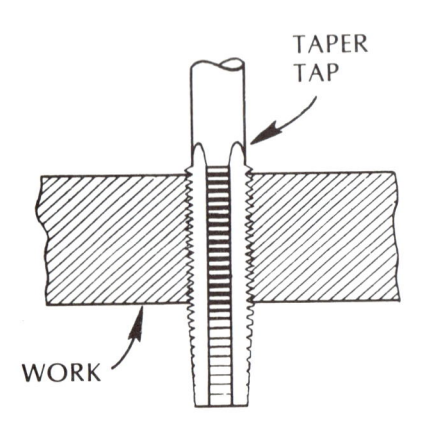

Fig. 5-43: Tapping through work with a taper tap.

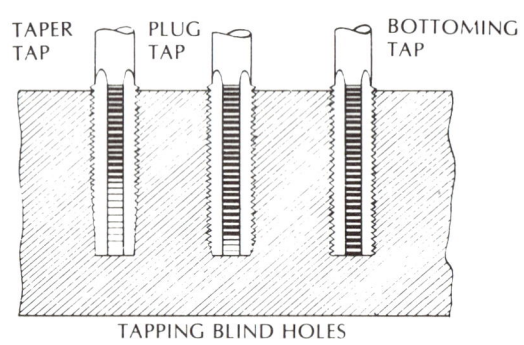

Fig. 5-44: A blind hole is tapped with three kinds of taps in succession.

Chapter 6

SANDING ON THE DRILL PRESS

A wide variety of sanding, polishing, and buffing jobs can be done on the drill press, using sanding drums or disks, rotary rasps or files, and buffers. *When using any of these abrasive accessories make certain that the area is adequately ventilated.*

SANDING

One of the special uses of a drill press is for sanding the surfaces of wooden or metal parts. Sanding drums of various sizes ranging from about 11/16 inch to 3 inches in diameter are available. While smaller drums (those in the 11/16 inch size) may work through the hole in the drill table, the larger sized drums require an auxiliary table to utilize their full sanding surface (Fig. 6-1A). The simplest table can be made by mounting a 3/4 inch board on the drill table and cutting a hole, slightly larger than the drum, in the board. Run the drum down into the hole and clamp the quill so that the abrasive sleeve will extend just below the work and cut across the entire surface equally. An insert can be made so that the table will accommodate several different sizes of drum sanders (Fig. 6-1B). There are times when a specially sized drum would come in handy. To make such a drum, all you have to do is glue abrasive

Fig. 6-1: While working the hole in the table (A) the smaller drum sizes will give full surface exposure; the larger ones require an auxiliary table (B).

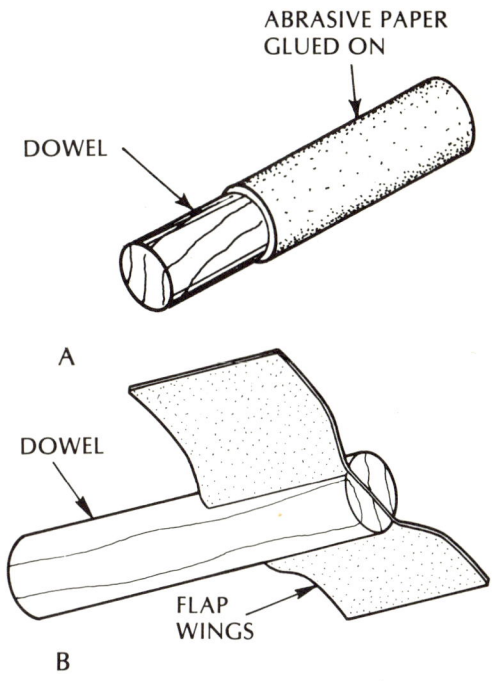

Fig. 6-2: Sanding drums you can make.

paper around a suitable size of dowel (Fig. 6-2A). The flexible sander, shown in Fig. 6-2B, is ideal for sanding odd-shaped workpieces or for smoothing small holes. It is made by inserting the abrasive paper in a slot cut in a dowel. Make sure this strip, which is two strips of abrasive paper glued back to back, is a tight fit in the dowel slot.

If you plan to do any large amounts of sanding on the drill press, the auxiliary table shown in Fig. 6-3 is excellent. The half-hole is made to take a 3-inch drum and will be satisfactory for all smaller drums. The auxiliary wood table shown in Fig. 6-4 is primarily intended for circular work, but thanks to the miter gauge slot, it can be used on straight jobs too.

Most sanding drums have a nut at one end, which, when tightened, expands the rubber drum to grip the sleeve. When the nut is loosened, the sleeve can be removed for replacement (Fig. 6-5). Garnet-coated sleeves are generally preferred for sanding wood, while aluminum oxide-coated sleeves are considered best for metals and plastics. Both types are usually available in coarse (40 grit), medium (50 grit) and fine (60 to 80 grit).

When sanding wood, high speeds with a fine abrasive tend to quickly clog the abrasive sleeve. Longer sleeve life is possible at lower speeds, while the best finish is obtained with a higher speed. A speed of about 1250 rpm is satisfactory for 3-inch drums. Drums of smaller diameters may be operated at slightly higher speeds. Avoid the use of excessive pressure which increases side thrust on the spindle and bearings. Abrasive sleeves remove material by the cutting action of the grit. Sanding operations will be effective if the proper drum speed and the correct type and size of grit for the material being worked are used.

Contour Sanding. When sanding on a drum sander, always feed the work *against* the rotation of the drum for reasons of safety (Fig. 6-6). Most curved work is done freehand since it is possible to move and turn the workpiece to suit the contour (Fig. 6-7). Keep a firm grip on the work and move it along steadily, because resting it in one place will result in an indentation and burning.

One problem frequently encountered is the fitting of a leg to a round column.

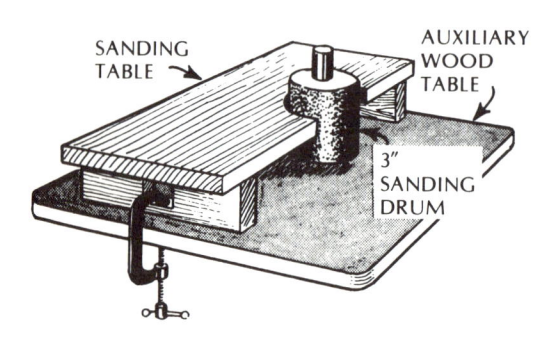

Fig. 6-3: Elevated table allows use of full length of sanding drum.

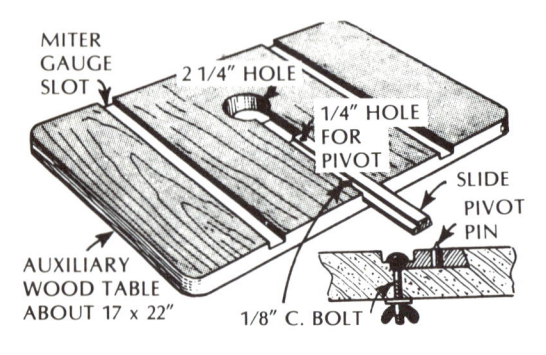

Fig. 6-4: A wood sanding table with pivot pin for circular work.

Fig. 6-5: The basic drum sander assembly.

Fig. 6-6: Feed workpiece against the rotation of drum.

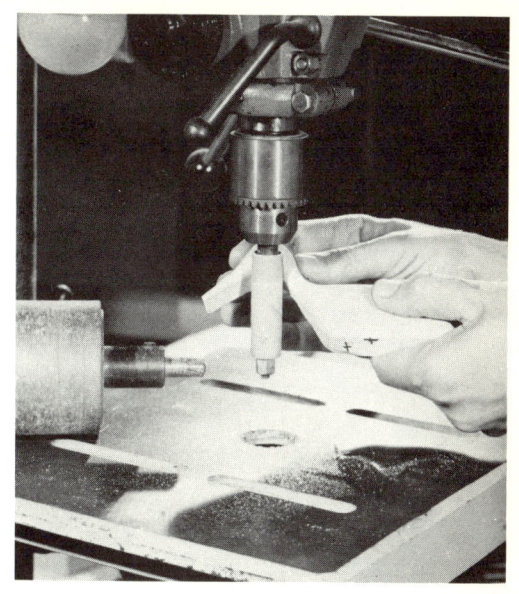

Fig. 6-7: Freehand sanding on a sanding drum.

Figure 6-8 shows how a drum sander setup in the drill press can be used to solve the problem, if you have a tilting table. The diameter of the drum should be the same as that of the column to which the leg is to be joined. This setup requires the swinging of the drill press table into a vertical position, then clamping to it a piece of stock of suitable thickness, to bring the center of the drum in line with the center of the leg on which the concave surface is to be shaped.

Where accuracy of contour must be maintained, it is possible to do pattern sanding. The pattern rides against a wooden collar fitted to the drill table, Fig. 6-9, controlling the sanding cut on the work to which it is attached. The collar should be the same diameter as the sanding drum. Since only the bottom of the drum can be used, this works best with narrow face drums. Figure 6-10 shows preliminary steps in fitting the pattern with anchors and sawing the work to within 1/16 inch or less of the edge of the pattern. This method normally allows for unlimited workpiece size (Fig. 6-11).

Fig. 6-8: Concave sanding on a sanding drum.

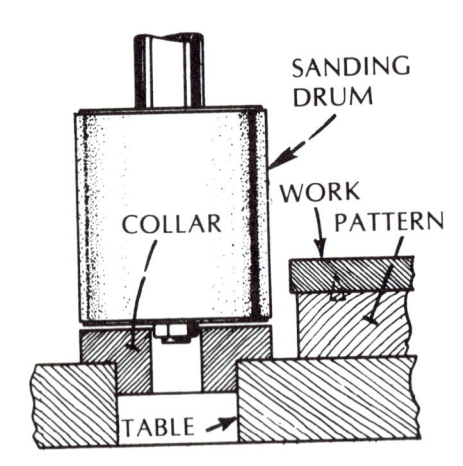

Fig. 6-9: The pattern sanding set-up.

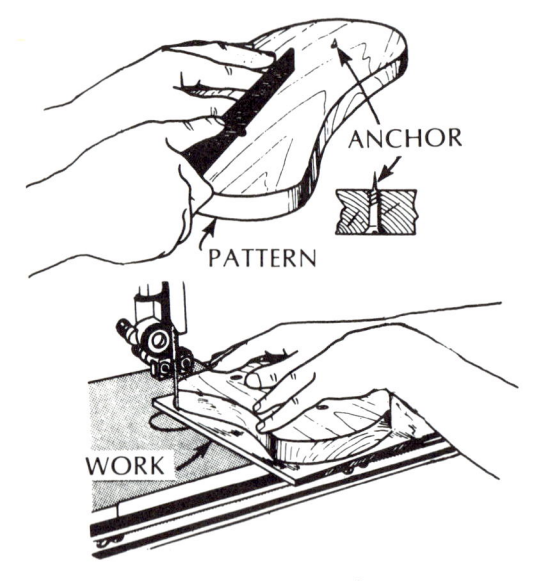

Fig. 6-10: How a pattern is made.

Fig. 6-11: Pattern sanding can be done on big work, too.

Straight sanding. Although nearly all curved drum sanding is done freehand, straight work usually requires a guide fence. The workpiece is run between the drum sander and the fence.

In sanding straight work, the work must be kept moving at a uniform rate past the drum (Fig. 6-12). If the work is stopped at any point while in contact with the rotating drum, it may be scored or burned. Uneven feed can produce scoring at intervals along the length of the stock. On short work, such as that pictured in Fig. 6-13, it is possible to make the sanding stroke with one sweep of the arms without removing either hand from the work. On longer stock it will be necessary to shift the hands alternately. Here, the trick is to maintain uniform pressure and rate of feed with one hand while the other is being shifted. Do not try to remove a great deal of material in one pass. Angular sanding can be done by tilting the table (Fig. 6-14), or by using an auxiliary tilting table. The head of a radial drill press can be set at any angle to accomplish angular sanding. Freehand sanding is often done with the table swung to one side. As shown in Fig. 6-15, it can be used as a support when in this position.

If you use an auxiliary table such as is shown in Fig. 6-4, making circles is no problem (Fig. 6-16). Advance the work until the drum starts cutting. Then lock the slide and rotate the work against the rotation of the drum. To sand small circular

Fig. 6-12: Feeding past a drum sander.

Fig. 6-14: Angular sanding accomplished by tilting the table.

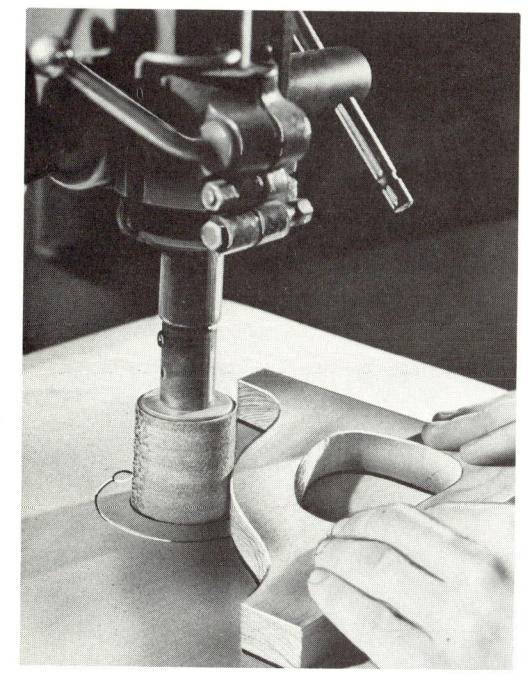

Fig. 6-13: Small work can also be fed past the sander.

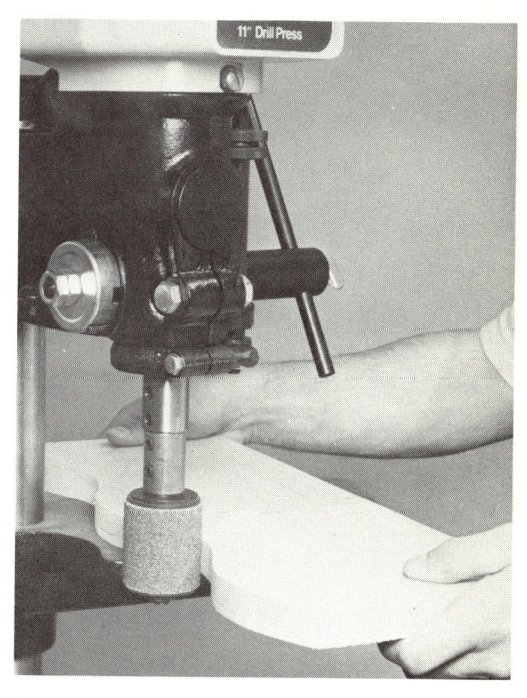

Fig. 6-15: Freehand sanding can be done by moving the table to one side.

101

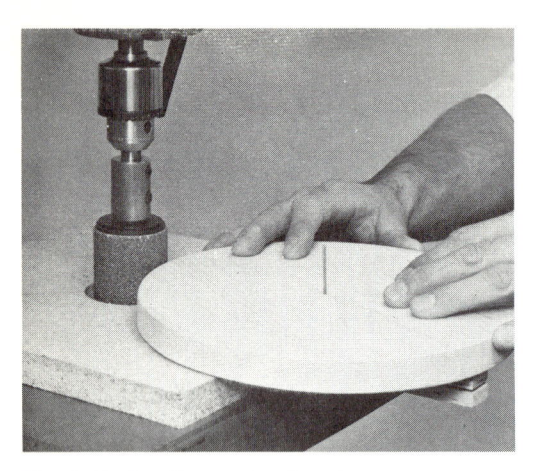

Fig. 6-16: Sanding a circular piece.

pieces, like dowels, that can be held in the chuck, drill a hole in the auxiliary table, so that the free end of the workpiece can fit loosely into it (Fig. 6-17). This arrangement will prevent the work from whipping.

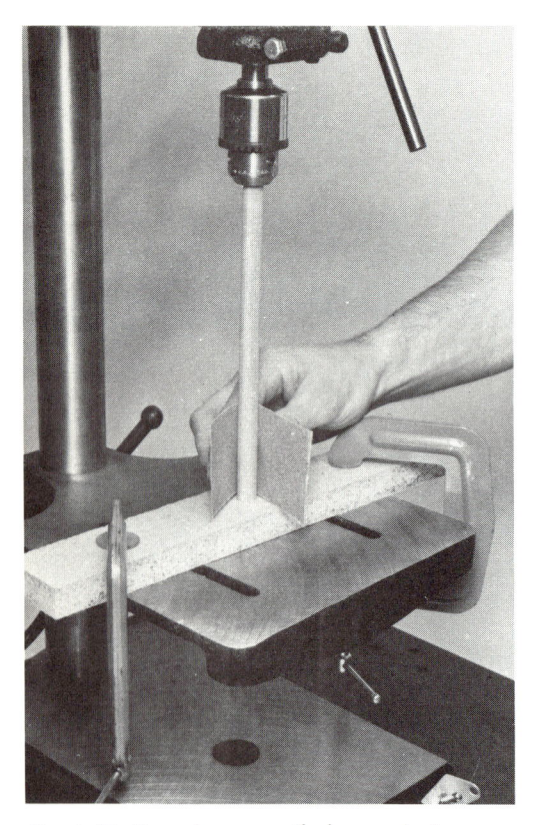

Fig. 6-17: Dowels can easily be sanded.

Horizontal Operation with a Radial Drill Press. While most drum sanding operations with a standard press are achieved with its spindle in a vertical position, the radial drill press also permits use as a horizontal drum sander. In this position, it does an effective job of surfacing narrow work. For this operation, use an auxiliary table and set the spindle in a horizontal position. Place the workpiece tightly against the fence and feed it against the rotation of the drum.

Wider boards may be handled in the same manner except that several passes will have to be taken with the sander at the same height. Remember that in any surface sanding operation you should not attempt too deep a bite in one pass; two or more passes will result in a better job.

Fig. 6-18: Sanding a rabbet with a drum sander.

Sanding Rabbets and Similar Cuts. Sanding the inside corners of rabbets and similar cuts can be executed easily with the drum sander in a horizontal position as shown in Fig. 6-18. The rabbeted stock is set against the auxiliary table guide fence, and the drum is set to fit in the corner. Then the work is fed forward past the drum to make the cut. For operations like this, the sleeve should be mounted so that it projects about 1/32 inch beyond the bottom of the drum, allowing the inside corner to be finished cleanly.

The radial drill press is also handy for smoothing miter cuts, and for freehand work such as that shown in Fig. 6-19.

Fig. 6-19: Freehand sanding on a radial drill press.

Disk Sanding. As mentioned in Chapter 1, the sanding disk kits sold for use with electric drills can also be employed on a radial drill press or standard type drill press with a tilting table. On small work the auxiliary tilting table may also be used. When disk sanding, never apply the disk flat on the work. If you keep the disk in one spot, you will cut only circular grooves. The abrasive will also clog very rapidly and the surface could become damaged.

The tilt of the table on the radial head should be such that approximately 1/4 or 1/3 of the abrasive surface comes in contact with the workpiece. Apply just enough pressure to bend the rubber backup pad. Also start the drill press motor before making contact with the work and keep the workpiece moving continuously (Fig. 6-20). Work with the grain when doing the final strokes. Disk sanding can also be done on a standard drill press by either tilting the table or by using a riser block as shown in Fig. 6-21.

Sanding Metals and Plastics. Finishing metals and plastics with either a drum or a disk sander on a drill press is the same as similar operations on wood, with the exception of the abrasive used. With these materials the feed should be less and the pass made more slowly.

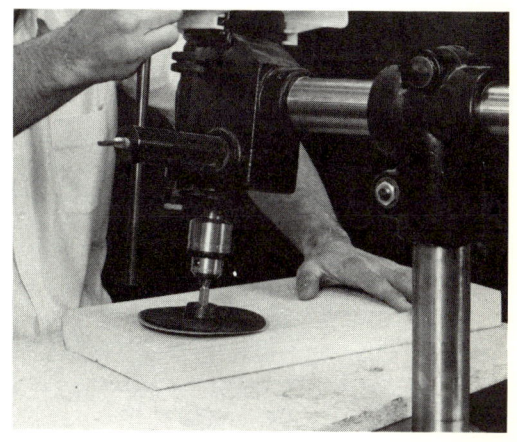

Fig. 6-20: Disk sanding on a radial drill press.

Fig. 6-21: Using a riser block when disk sanding.

103

ROTARY RASPS AND FILES

Rotary rasps, although primarily woodworking tools, find wide usage on soft metals, rubber, fiberglass, plastics and many composition materials. These versatile tools are ideal for fast stock removal; for roughing, finishing, shaping, and profiling; for elongating holes and slots; and for roughing prior to bonding rubber. The disk rasp can be used in the same manner as when doing disk sanding. It will remove the material much faster, but it will leave a much rougher surface. The drum rasp (Fig. 6-22) has the same characteristics, but is used in a manner similar to that of the drum sander.

For most rotary rasp work, a spindle speed of 1250 rpm is generally best. Light feed pressure will keep the teeth sharp, longer. When shaping and cutting with a rotary rasp (Fig. 6-23), treat it like a drill, retracting it frequently to remove waste. Also do not attempt to cut too deeply. A typical one pass cut is a groove approximately 1/4 by 3/8 inch deep in pine. Keep the workpiece flat on the table with the quill locked at the desired depth. If the piece is too small to hold securely, clamp it in a vise. Do not attempt to hold the workpiece in both hands. You cannot properly steady it this way.

While rotary files are used basically as metal working tools, they find wide usage on plastics, hard rubber, wood, and many building materials. These versatile tools can file and countersink; can elongate holes and slots (Fig. 6-24); can remove burrs and scale; and can perform profiling, light milling, finishing and blending operations. Operate the rotary file at slow speeds and be sure to *always wear safety goggles.*

With a flexible shaft in your chuck (Fig. 6-25), you can drill, sand or do any of the various rotary rasp and file operations away from your drill table. However, make sure not to overload the flexible shaft. With a drill press, the available torque is usually beyond the capacity of the shaft. Let the

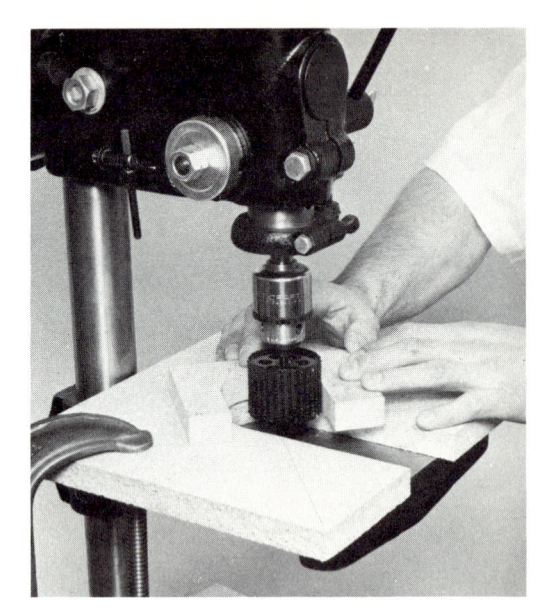

Fig. 6-22: A rotary drum rasp in use.

Fig. 6-23: Cutting a rabbet with a rotary wood rasp.

shaft, not the source of power, determine the capacity. Do not operate the flexible shaft at less than 1250 rpms nor more than 2400 rpms. The shaft should not be bent through an angle smaller than 90 degrees or in a radius less than 4 inches. The shaft should be lubricated as directed by the manufacturer.

BUFFING ON A DRILL PRESS

The process of restoring a gleaming finish to a pitted or badly tarnished metal surface can be performed easily on your drill press. To buff a metal surface, mount a cloth buffing wheel and arbor on the drill press spindle.

Fig. 6-24: Elongating a hole with a rotary file.

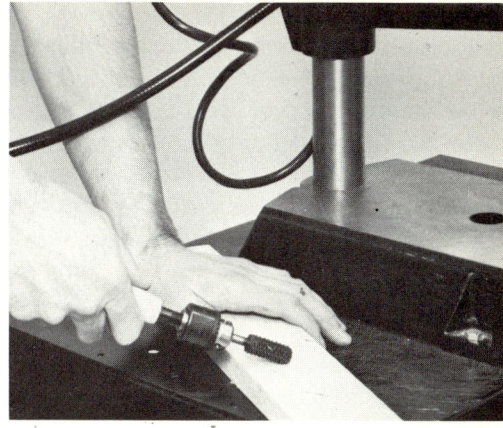

Fig. 6-25: A flexible shaft has many uses.

If the object is badly pitted, you will have to cut the entire surface down to the level of the deepest pit, using an emery composition which comes in a tube or stick. To apply the emery to the buffing wheel, hold the tube to the face of the running buffer and let the heat of friction melt the binder so that it will flow onto the buffing wheel. The face of the wheel will turn black when it is well charged.

After the entire metal surface is uniformly clean, put on another wheel and charge it with tripoli, a much finer abrasive. Go over the metal very lightly, using downward strokes. The scratch lines left by the emery polishing will smooth out and blend into each other.

Although the object will have a good finish now, it is usually desirable to give it a really fine finish by a final buffing, or coloring, with a buffer wheel charged with jeweler's rouge (Fig. 6-26). Final color buffing will show up imperfect work in the preliminary stages. If you find such spots, go back over them with the tripoli-coated wheel.

Buffing wheels charged with suitable compounds are used for polishing bare wood, lacquered surfaces, plastic, and so on. Such easily obtained abrasives as pumice, rotten stone, and rouge will usually do the job. Mineral spirits or wood alcohol will remove any film from the buffing compound left on the work after buffing.

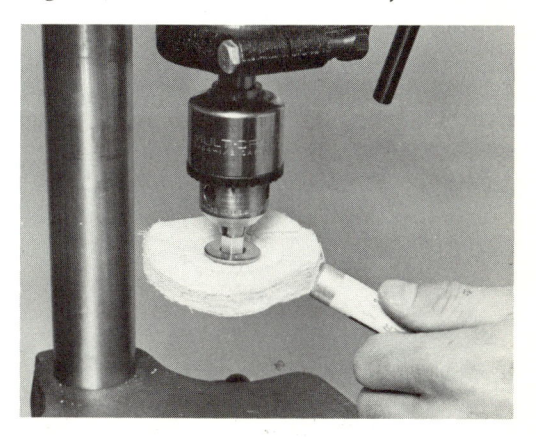

Fig. 6-26: Charging a buffer wheel with jeweler's rouge.

WOOD AND METAL POLISHING

To convert the drill press into a polishing machine suitable for such operatons as polishing wood and metal surfaces, all that is required is a sanding disk and a lamb's wool bonnet (usually sold as a kit for electric hand drills).

Secure the lamb's wool bonnet to the metal disk plate by means of a drawstring provided in the bonnet. Then mount the plate in the chuck and lock the quill at a convenient height. Use a 1250- to 2400-rpm spindle speed and *protect your eyes with goggles.*

Feed the work to the bottom side of the

bonnet (never the edge), as shown in Fig. 6-27. Coat the workpiece surface with wax, oil, or another similar polish. Always employ a light feed pressure.

Metal rods can be easily polished by inserting them in the drill press and applying an abrasive strip as shown in Fig. 6-28. To get the best results, work through progressively finer abrasive strips, ending by using fine steel wool.

DAMASKEENING METAL

The damaskeen or spot finish is worked on metal and consists of a pattern of overlapping rings. This work requires some practice to get good results. Two methods can be used. With the first, a hole of the desired spot size is drilled in a piece of thin, hard brass. This is used as a template, hold-down, and container for the abrasive slurry. The abrader is simply a short length of dowel (Fig. 6-29), while the abrasive can be 150-grit emery powder, aluminum oxide, or silicon carbide grains. The abrasive is fed to the work mixed with oil or water. The second method uses an abrasive disk of the required size. This must be fitted to cork (Fig. 6-30) or live rubber since a certain amount of flex is needed. Instead of using an abrasive disk, it is practical to feed a slush of abrasive powder under a revolving cork. The eraser of an ordinary pencil gives good results with this method. Whatever the method, overlap the rings as desired for a suitable pattern. The work can be done with a freehand design or may be controlled with any style of spacing device. Operate the spindle at 1250 rpm.

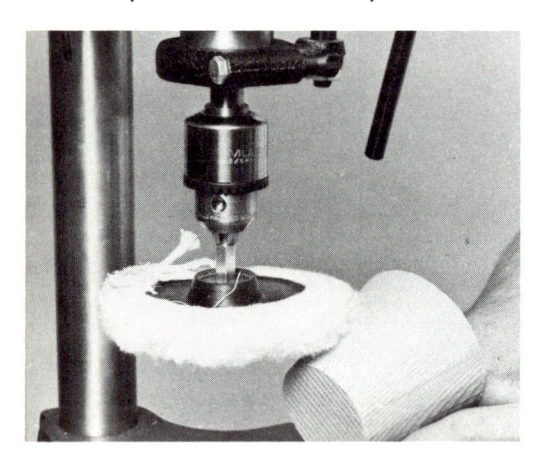

Fig. 6-27: Proper way of polishing.

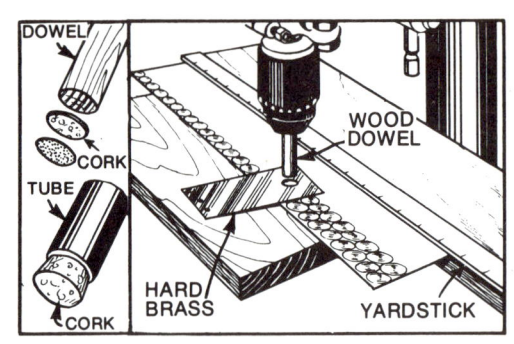

Fig. 6-29: Damaskeening technique.

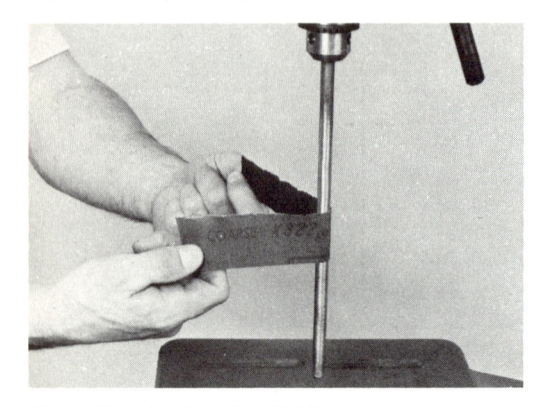

Fig. 6-28: Metal rod polishing.

Fig. 6-30: Damaskeening with a cork and tube.

Chapter 7

OTHER OPERATIONS ON THE DRILL PRESS

There are many additional jobs you can do on a drill press. Additional operations range from spinning rivets, polishing plastic balls, to drilling such substances as plastic, hard rubber, glass, and various composition materials. While the basic drilling techniques for these latter materials are the same as for metal, however, there are a few special procedures that should be discussed.

DRILLING PLASTIC AND HARD RUBBER

While both plastic and hard rubber (Fig. 7-1) can be drilled with ordinary metal twist drills, better results are achieved if the bit points are reground to the lip angle illustrated in Fig. 8-10. This is especially true in the case of some of the softer plastics in which a standard drill may tend to dig into the material and take excessive bites. Full details on how to regrind the lip angle are given in Chapter 8.

Fig. 7-1: Drilling holes in rubber. Sometimes a water coolant is needed to keep the drill from overheating.

The same spindle speeds can be used for both materials (see page 40). Lift the bit out of the hole frequently to clear the chips and prevent the softening or burning of the material, particularly when drilling a thermosetting plastic. If the holes are for metal (screws, nails, and so on) and if the material will be subject to temperature changes, it is a good idea to drill them 1/64 to 1/32 inch oversize to allow for expansion and contraction.

DRILLING GLASS

The drilling of glass is an operation that may be done easily on the drill press, although it is difficult by any other method. Actually there are two methods of drilling glass on a drill press. The first employs either a diamond-pointed drill bit or a specially designed tungsten-carbide pointed drill bit (usually available at jeweler's supply houses and some specialty hobby shops). The glass workpiece must be clean and well supported on a flat piece of rubber, felt, carpeting, or wood. If the latter is used, it must be *absolutely* smooth and level (the slightest bump—even a raised grain—under the pressure point can start a crack).

Use the slowest possible spindle speed and keep the drill bit well lubricated; kerosene, camphor oil, or turpentine is best, though water will do. A light but firm, even feed pressure is needed to keep the bit grinding (there are no visible chips.) Heavy feed pressure heats the glass and invariably causes breakage. In addition, the pressure should be eased off quickly when the bit point breaks through. Better, if practical, is to turn the work over and finish from the other side at this point.

Because the diamond-pointed bit and the glass cutting tungsten-carbide pointed bit are expensive, the second method of drilling glass is often used. The drill employed in this procedure is a piece of brass tubing with an outside diameter equal to the size of the hole to be drilled (Fig. 7-2). The tubing should be slotted with one cut, using a very narrow saw. The cut need not extend more than about 1/4 inch from the end of the tube. Similar results are effected by notching the end of the tube in two or three places. The tube is not sharpened in any way; it is simply cut square on the end and then slotted or notched as noted. The glass should be supported on an *extremely* flat piece of wood or, better, on a piece of felt, carpeting or rubber. A dam of putty is built around the area where the hole is to be drilled (Fig. 7-3), or a felt ring can be used for the same purpose of keeping the lubricant and cutting compound in place. The well is fed with a fluid mixture of 80- or 100-grit silicon-carbide abrasive powder combined with camphor oil or turpentine. Use the slowest possible spindle speed and employ only enough feed pressure to keep the bit grinding. Heavy pressure must be avoided as this will invariably break the glass—15 pounds of pressure is sufficient. Since the leverage gained by a rack-and-pinion drill feed is about 15:1, you only need about 1 pound of pressure on the lever. A 1-pound weight on the feed lever will help to maintain the required pressure. Do not attempt to drill completely through; about halfway through, restart the hole from the other side.

For very small holes, the tapering triangular point is useful. This can be ground from a slim, three-corner file, bringing the point to needle sharpness and then grinding the end off square, to make a flat triangle about 1/32 inch wide. Work this with light pressure, using turpentine or camphor oil as a lubricant.

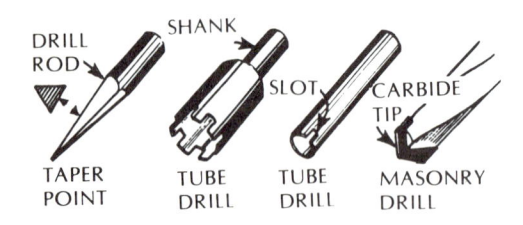

Fig. 7-2: Tools used in drilling glass.

Fig. 7-3: Putty dam used in drilling glass.

Fig. 7-4: Drilling a ceramic object using turpentine and an abrasive powder.

Clay pottery and hard glazed ceramic surfaces (Fig. 7-4) such as tiles can be drilled in the same manner as glass.

Large-diameter holes in asbestos board and similar composition materials can be cleanly drilled with a hole saw.

MIXING PAINT

Paint mixing paddles can be purchased (Fig. 7-6) or made as is shown in Fig. 7-7. If a full can of paint is being mixed, the pad-

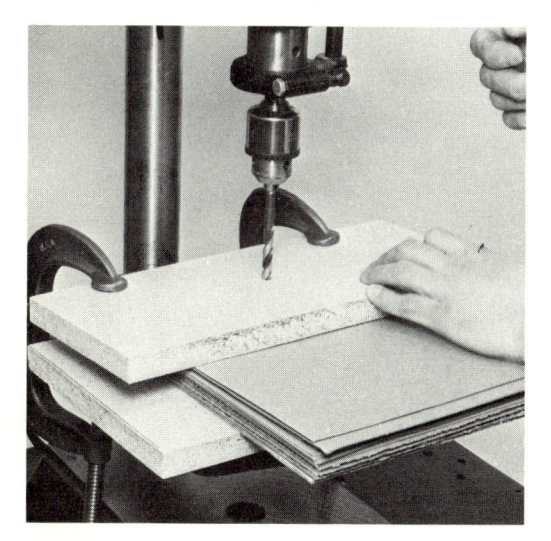

Fig. 7-5: Set-up for drilling paper. Be sure the area is well ventilated.

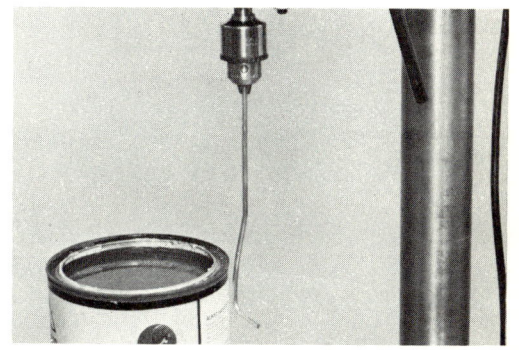

Fig. 7-6: Using a ready-made paint stirrer.

DRILLING PAPER

Spur bits will do a perfect job of drilling holes in paper if the work is tightly clamped between boards, as shown in Fig. 7-5. The speed should be about 1200 rpm. The bit should be lifted frequently to clear the rings of paper which will collect on the tip of the drill.

DRILLING PLASTER AND OTHER SIMILAR MATERIALS

Drilling materials containing gritty or abrasive substances like some types of wall board will very quickly dull a tool-steel twist drill. If a large amount of these materials must be drilled, use a tungsten-carbide-tipped bit. Many of these products are covered with a paperlike material.

To make a clean hole in any soft-bonded or paperlike material, use the highest possible spindle speed and a light feed pressure to prevent the bit from tearing the material. If possible, drill only halfway through (or until the drill bit tip just penetrates); then turn the work over to finish from the other side.

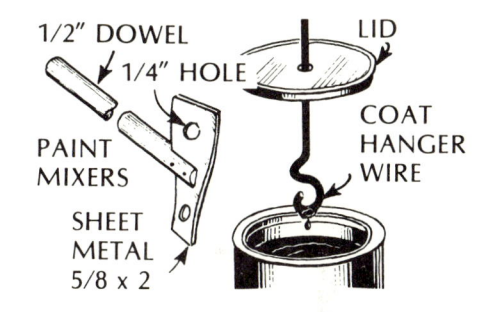

Fig. 7-7: Paint mixers you can make.

dle should work through a hole in the lid. The preferable practice is to pour off some of the oil. Then, with the drill running at a slow speed, any type of small paddle can be used without having the material splash out of the can. Do not allow the paddle to hit the edge of the can. When mixing paint, be sure that sufficient ventilation exists.

SPINNING RIVETS

Hollow metal rivets, used extensively for fastening cardboard, sheet metal, and other materials, can be successfully worked on the drill press by a spinning operation.

As shown in Fig. 7-8, a vise and special rivet set are necessary. The vise is bolted to the drill press table, and must be centered exactly with the setting tool which is held in the chuck. The rivet is inserted through the work, placed on the vise, and the rivet set is then brought down to turn the rivet over to a neat finished edge. A speed of about 2400 rpm should be used. A higher speed is acceptable providing the rivet set is properly hardened.

MAKING FLAT DRILLS

Occasionally, when an odd-sized drill is needed the knowledge of how to make flat drills will prove valuable. These are made from lengths of drill rod. To make the drill, cut off a short length of drill rod of suitable diameter; then forge or hammer the end down in a long taper, as shown in Fig. 7-9. Flatten the end of the tapered part until it is about 1/16 inch thick at the point. To temper, heat the end to a bright cherry red; then plunge about 1 inch of the end into tepid water. Quickly brighten the face of the flat with a piece of emery cloth on a stick, and watch the colors run up the point from the still hot portion of the drill. When the point is a light straw-yellow color, immediately quench the whole drill in water. Grind the point to the shape shown, the angle of the point being the same as for twist drills, 59 degrees with the center. Rake and clearance on the point and sides should be ground as indicated. Extreme care must be taken when working with heated metals.

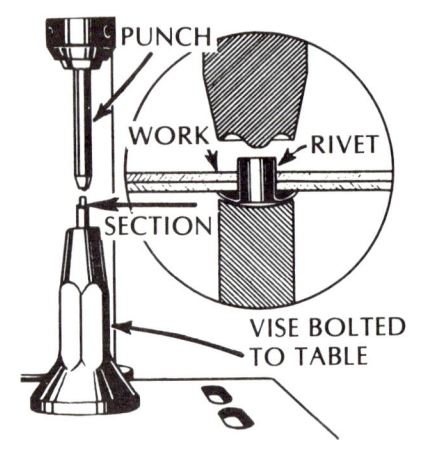

Fig. 7-8: Method of spinning rivets.

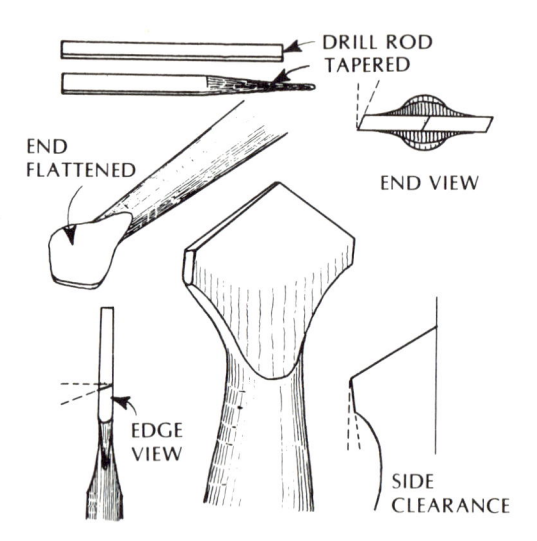

Fig. 7-9: How to make your own flat drills from drill rod.

POINTING DOWEL RODS

The pointing or beveling of wood or plastic rods can be done on the drill press by pressing the revolving work against a file or piece of sandpaper held at the required angle in a suitable jig block (Fig. 7-10). While cutting, the holder should be moved back and forth to keep the dowel from ridging the guiding edge.

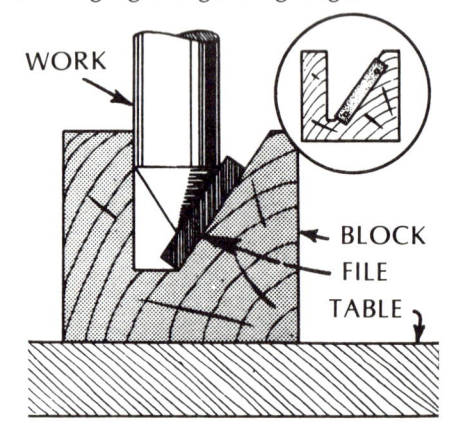

Fig. 7-10: Method of pointing dowel rods.

Chapter 8

SHARPENING DRILLS AND BITS

To get the most out of your drill press, drills, bits, and other cutting tools must be kept sharp. For example, as mentioned previously, the twist drill can be rated one of the most efficient of all cutting tools, based on the length of the cutting edge in proportion to the amount of metal which supports it. It is a tool which will stand a tremendous amount of abuse and still keep cutting. However, any drill will work better and last longer if it is properly ground. Grinding is a simple operation, easily mastered by the average person after a brief study of the fundamental mechanics of a good cutting edge.

SHARPENING TWIST DRILLS

As mentioned in Chapter 1, the two most important features of drill-grinding are (1) the point angle, and (2) the lip clearance. In addition, you must consider the correct clearance behind the cutting lips and the correct chisel-edge angle. All four play an important part when grinding either a regular point (Fig. 8-1), which is used for general purposes, or a flat point (Fig. 8-2), which is used for drilling hard, tough materials.

It is easy to check the drill point. Gauges in a variety of styles can be purchased at a nominal cost, or can be made from sheet

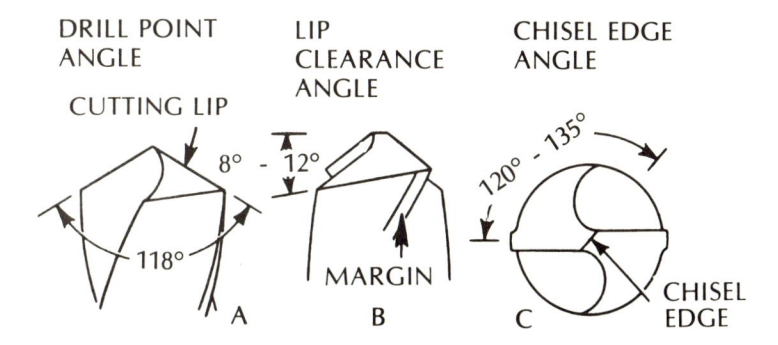

Fig. 8-1: Specifications for grinding a regular point twist drill.

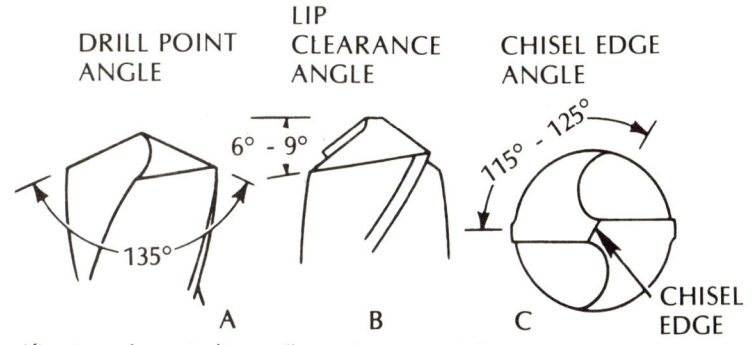

Fig. 8-2: Specifications for grinding a flat point twist drill.

metal. Figure 8-3 shows a drill being checked during grinding. The drill-point gauge is being held against the body of the drill, and has been brought down to where the graduated edge of the gauge is in contact with one cutting edge. In this way, both the drill-point angle and the length of the cutting edge (or lip) are checked at the same time. The process is repeated for the other side of the drill.

Lip clearance behind the cutting lip at the margin is determined by inspection. This means that you look at the drill point and approximate the lip-clearance angle (see Figs. 8-1B and 8-2B), or compare it to the same angle that has been set on a protractor. The lip-clearance angle is not necessarily a definite angle, but must be within certain limits. Notice that in Fig. 8-1B this angle ranges from 8- to 12-degrees, and that the range given in Fig. 8-2B is 6-to 9-degrees. Whatever angle in the range is used, however, lip clearance should be the same for both cutting lips of the drill.

There must be lip clearance behind the entire length of the cutting lip which extends from the margin of the drill to the chisel edge. This means there must be "relief" behind the cutting lip along its entire length.

When lip clearance is being "ground into" a drill, the lip-clearance angle and the chisel edge angle (shown at C in Fig. 8-1 and 8-2) will be your guide to the amount of clearance you have ground into the drill behind the cutting lip along its entire length. The greater these angles are, the more clearance there will be behind the respective ends of the cutting lip. Too much lip clearance, which occurs when both the lip-clearance angle and the chisel-edge angle exceed their top limits, weakens the cutting edge or lip by removing too much metal from directly behind it. Too little or no lip clearance prevents the cutting edge from producing a chip or cutting, and the drill will not drill a hole.

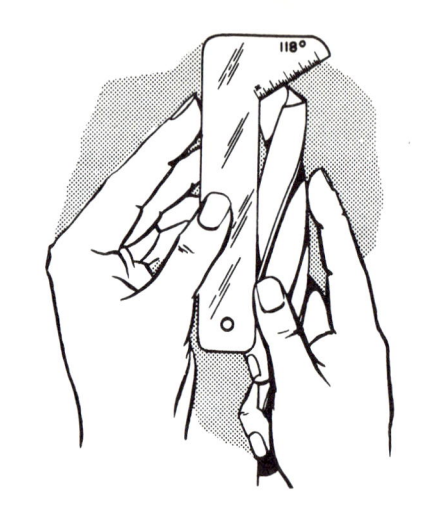

Fig. 8-3: Checking the drill point angle and cutting edge with a gauge.

Drill Grinding. To sharpen twist drills first get the grinder ready. If necessary, dress the face of the wheel so that it is clean, a true circle, and square with the sides. Before starting the grinder, readjust the tool rest to 1/16 inch or less from the face of the wheel. This is an important safety measure which will help keep work from wedging between the tool rest and the face of the wheel.

After starting the grinder and letting it come up to speed, you can begin grinding the drill point. Hold the twist drill as shown in Fig. 8-4A, which is a top view of the first step in grinding a drill. The axis of the drill in the first step, should make an angle of about 59 degrees (half of the drill-point angle) with the face of the wheel as shown in Fig. 8-5A. The cutting lip should be horizontal. The actual grinding of the drill point consists of three definite motions of the shank of the drill while the point is held lightly against the rotating wheel. These three motions are: (1) to the left, (2) clockwise rotation, (3) and downward.

Figure 8-5 shows the motion to the left in three views as the angle between the face of the wheel and the drill decreases from about 59 degrees to about 50 degrees.

The clockwise rotation is indicated by

A

B

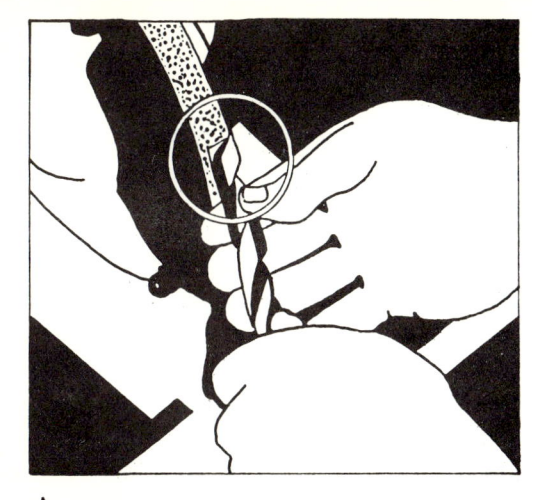

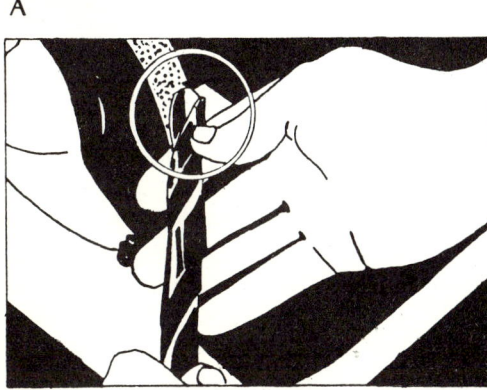

Fig. 8-4: (A) Grinding a twist drill with a grinder (initial position). (B) Grinding a twist drill with a grinder (final position). (**Note:***In all grinding operations be sure to wear safety goggles.*)

the advance of the rotation arrows in A, B, and C. Rotation is also pictured by the change in position of the cutting lip. The same operation is performed on the other cutting lip. Check to ensure equal cutting lip length and a proper point angle.

To aid in the grinding operation, a guide board can be made to fit on the tool rest of the grinder. Figure 8-6 shows how the guide can be laid out with scribed lines at 59 degrees and 47 degrees. These lines represent the positions during the grinding rotation of the drill.

Web-Thinning. Repeated sharpening, which shortens the drill, results in an increase in the web thickness at the point. This may require web thinning. Proper web thinning, when it becomes necessary, is important for satisfactory drilling.

To thin the web of a drill, hold the drill lightly to the face of a round-edge wheel, as shown in Fig. 8-7A, and thin the web for a short distance behind the cutting lip and into the flutes. This is shown in Fig. 8-7B.

A

B

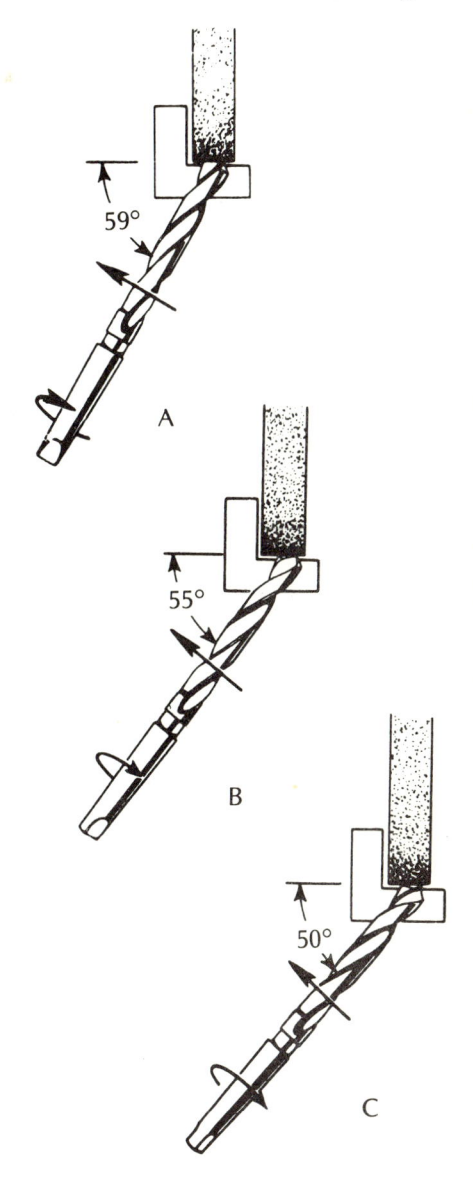

C

Fig. 8-5: Three steps for grinding a twist drill with a grinder.

Notice that the cutting lip is actually (but only slightly) ground back, reducing its included angle only a little and not enough to affect the operation of the drill.

If you do not have a suitable round-edge wheel, the web can be thinned on an ordinary square-face wheel. In this

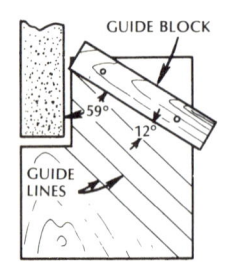

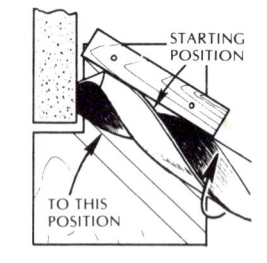

Fig. 8-6: Grinding guide.

method of web-thinning, the grinding is done on the back of the lips, the grind being carried up to the center of the point on each side, as shown in Fig. 8-8.

Drill Point Variations. While a point angle of 59 degrees and a clearance angle of from 12 to 15 degrees has been found best for average work, the greatest efficiency is obtained if drills are specially-ground for the work to be done. For example, fiber takes a drill with a point angle of only 30 degrees, while manganese steel requires a 75-degree point angle. Figure 8-9 shows the best drill points for various materials.

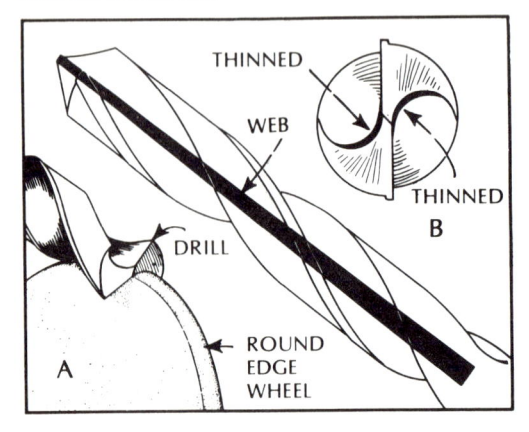

Fig. 8-7: Thinning the web on round-edge wheel.

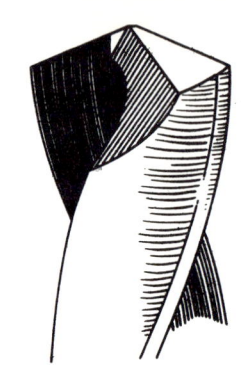

Fig. 8-8: Results of web thinning on a square-edge wheel.

For drilling brass, copper, and most plastics, it is advantageous to modify the cutting edges of the drill. The effect to be obtained is shown exaggerated somewhat in Fig. 8-10A, which shows how the cutting edge of the lip is ground off. This makes the edge scrape rather than cut, and reduces the tendency of the drill to "dig in" in brass and other soft metals. This form of grinding can be done on a fine-grit wheel, as shown in Fig. 8-10B. Very little metal is ground off—just a few thousandths of an inch—and for this reason it is often better to flatten the cutting edge with a small sharpening stone (Fig. 8-10C).

Wheels for Drill Grinding. The purified form of aluminum oxide (white in color) makes the best wheel for grinding high-speed drills. It is not as tough as regular aluminum oxide, but runs cooler and is practically self-dressing. A second choice for high-speed drills and a first choice for general drill grinding is the regular type of aluminum oxide. In either case, the grit should be about No. 60 and the grade medium hard. The best wheel shape is a recessed center, but good work can be done on the side of a straight wheel.

Commercial Grinding Attachments. Sharpening a twist drill by hand is a skill that is mastered after much practice and careful attention to details. Whenever possible, use a tool grinder in which the drills can be properly positioned, clamped in

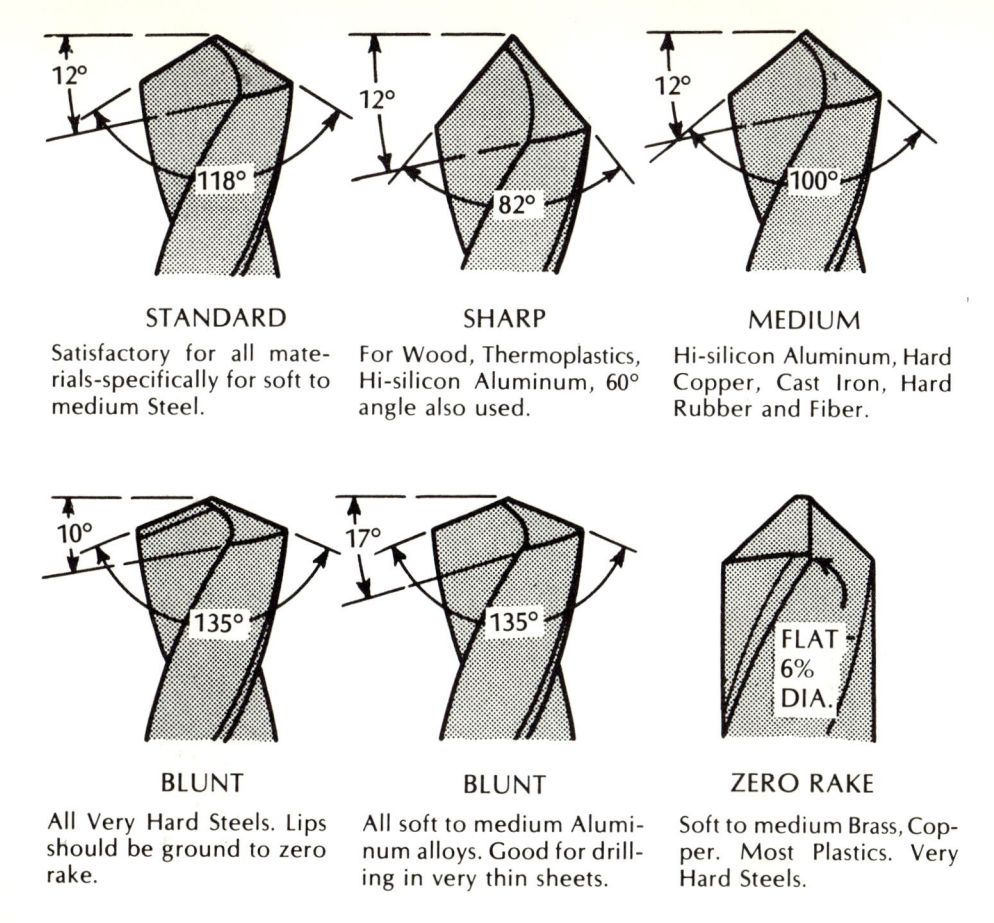

STANDARD
Satisfactory for all materials-specifically for soft to medium Steel.

SHARP
For Wood, Thermoplastics, Hi-silicon Aluminum, 60° angle also used.

MEDIUM
Hi-silicon Aluminum, Hard Copper, Cast Iron, Hard Rubber and Fiber.

BLUNT
All Very Hard Steels. Lips should be ground to zero rake.

BLUNT
All soft to medium Aluminum alloys. Good for drilling in very thin sheets.

ZERO RAKE
Soft to medium Brass, Copper. Most Plastics. Very Hard Steels.

Fig. 8-9: Drill points for various materials.

place, and set with precision for the various angles. This machine grinding (Fig. 8-11) will enable you to sharpen the drills accurately. Drills will last longer as a result and will produce more accurate holes.

To do good work with the drill grinding attachment, it is necessary to have a true and smooth grinding wheel. This is of such importance for good work that a diamond dresser and holder (Fig. 8-12), is commonly sold as part of the drill grinding attachment. When dressing, the point of the dresser should be on or slightly below center (Fig. 8-13), and should be on a negative angle of about 10 degrees, which is fixed automatically by the holder. Pass the dresser rather quickly across the face to

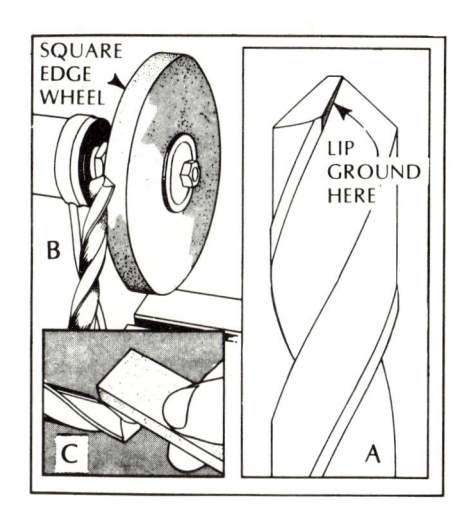

Fig. 8-10: Drilling in brass and other soft metal is best done with a drill ground to zero rake in the manner shown above.

115

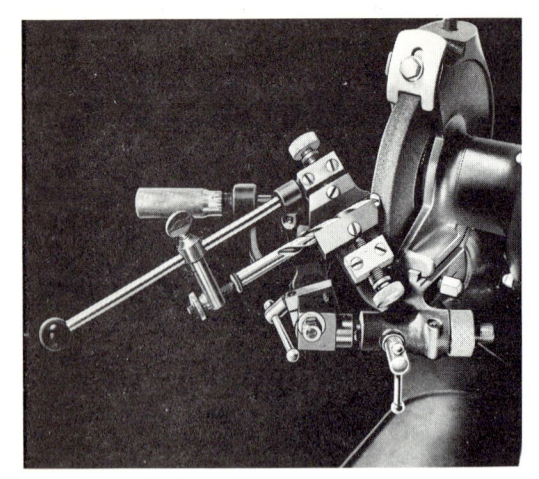

Fig. 8-11: Typical commercial grinding attachment.

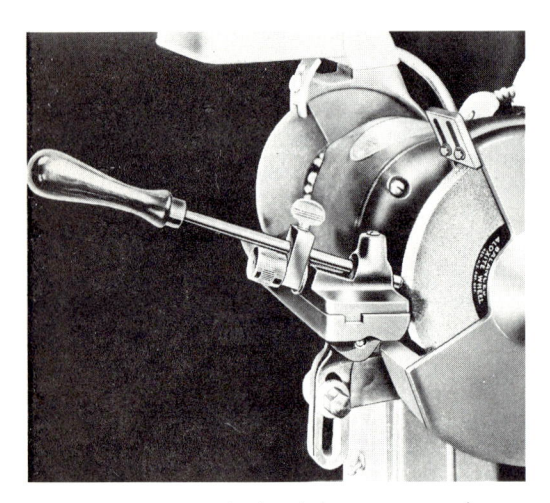

Fig. 8-12: Diamond wheel dresser at work.

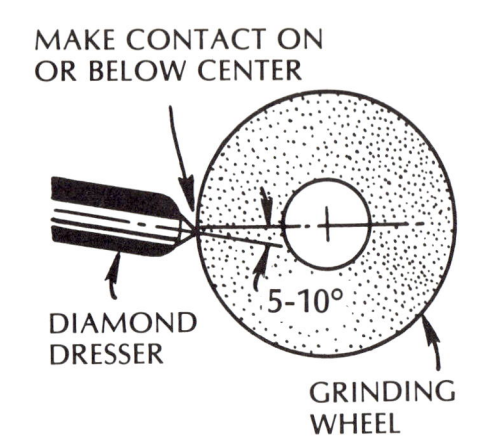

Fig. 8-13: Positioning on the wheel dresser.

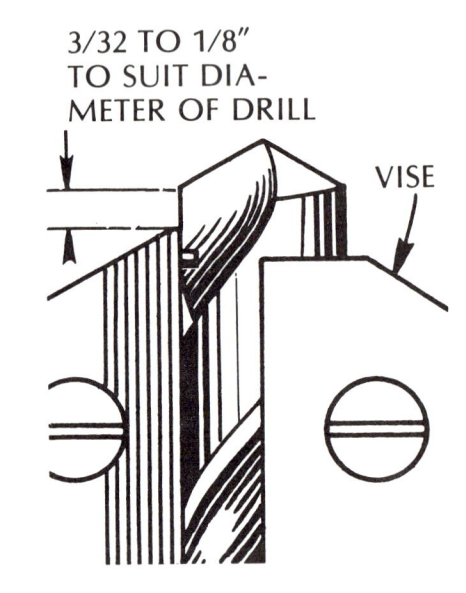

Fig. 8-14: Position of drill in a commercial grinder attachment.

keep the wheel open and free-cutting. Do not use more than .001 inch infeed per pass on the dresser.

One type of commercial drill grinding attachment is shown in Fig. 8-14. Instructions for its use can be found in the manufacturer's manual, which accompanies the machine.

Whether you are sharpening a drill by hand or by machine, it is very necessary that the temperature at the point be kept down. As the point gets hot, it approaches the temperature at which the temper of the steel will be drawn. Keep the point cool enough to be held in your bare hand. Do this by making a few light passes over the grinding wheel. Take a few seconds to let the point cool, and then repeat the alternate grinding and cooling.

When you notice the appearance of a blue color at the point, it is too late. You have drawn the temper and the steel is now too soft to hold a cutting edge. The only thing you can do is to continue the sharpening process, first one lip and then the other, until you have finally ground

116

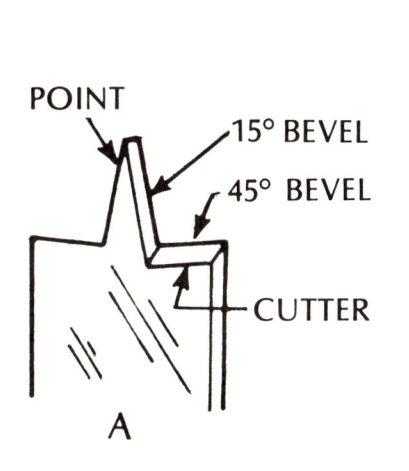

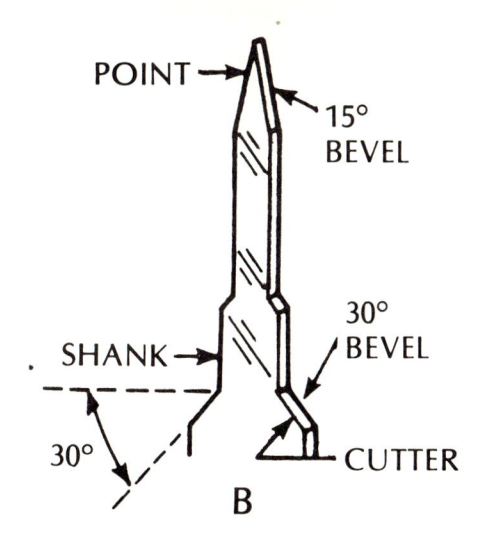

Fig. 8-15: Profiles of a speed bit (A) and a profile bit (B).

away the soft tip of the drill. This means that you must grind away all of that portion of the tip which is blue. As the blue color indicates softness throughout the entire point of the drill, and not only on the blue surface, resharpening must be continued until all of the blue-colored metal has been ground away. This operation must be done very slowly and carefully, keeping the point cool to prevent continued bluing of the metal.

SHARPENING SPEED OR SPADE BITS

When sharpening speed or spade wood-boring bits, use a medium-fine mill file with one edge ground smooth so that when filing the point of the cutter, the adjacent side will not be filed. Always maintain the original bevel (Fig. 8-15A). Since the two halves of the bit must be kept identical, remove as little metal as possible. Any wire edges are removed by honing on the flat side of a fine oilstone. Sharpen only the two cutting edges.

Screw profile bits, which are similar in shape to spade bits, are sharpened as above, except care must be taken not to file the shank. Maintain the 30-degree slope (Fig. 8-15B).

PLUG CUTTERS

When sharpening plug cutters, all grinding and filing should only be done on the outside, beveled rim and on the face of the cutting lip. The inside must never be touched except with a fine round oilstone, and then only to remove burrs that may have formed from grinding. When grinding the rim, maintain the original bevel angle.

WOOD AUGER BITS

Wood auger bits require hand sharpening. Before doing any sharpening, the spur must be straightened if it is out of shape. To do this, use a metal block which has a hole drilled large enough to receive the brad point of the bit. Position the bit so that the spur top rests over the edge of the metal block and then tap the spurs lightly with a hammer. After straightening the spurs, restore the outer contour with a file or by honing with a fine oilstone. Be careful, however, not to remove metal from inside the required circumference. The easiest way to do this is by touching an oilstone to the outer sides of the spur while spinning the bit on its center.

To sharpen the spurs, the necessary tools

are a small half-round file (Fig. 8-16A), an auger bit file (Fig. 8-16B), and a small square or triangular file. The two points which must be sharpened are the cutting lips and the cutting spurs. On wood bits of the twist drill pattern, touch up filing is done through the throat, as shown in Fig. 8-16C, using either the half-round file (if the throat is rounded) or the auger bit file (if the throat is open). Spurs are always sharpened on the inside, never on the outside (Fig. 8-16D). The top edge is filed as shown in Fig. 8-16E. The original bevel must be maintained. If only a portion of the lip is filed, the result will be as in Fig. 8-16F. With this type of angle the chip lifting ability of the cutting edge is destroyed.

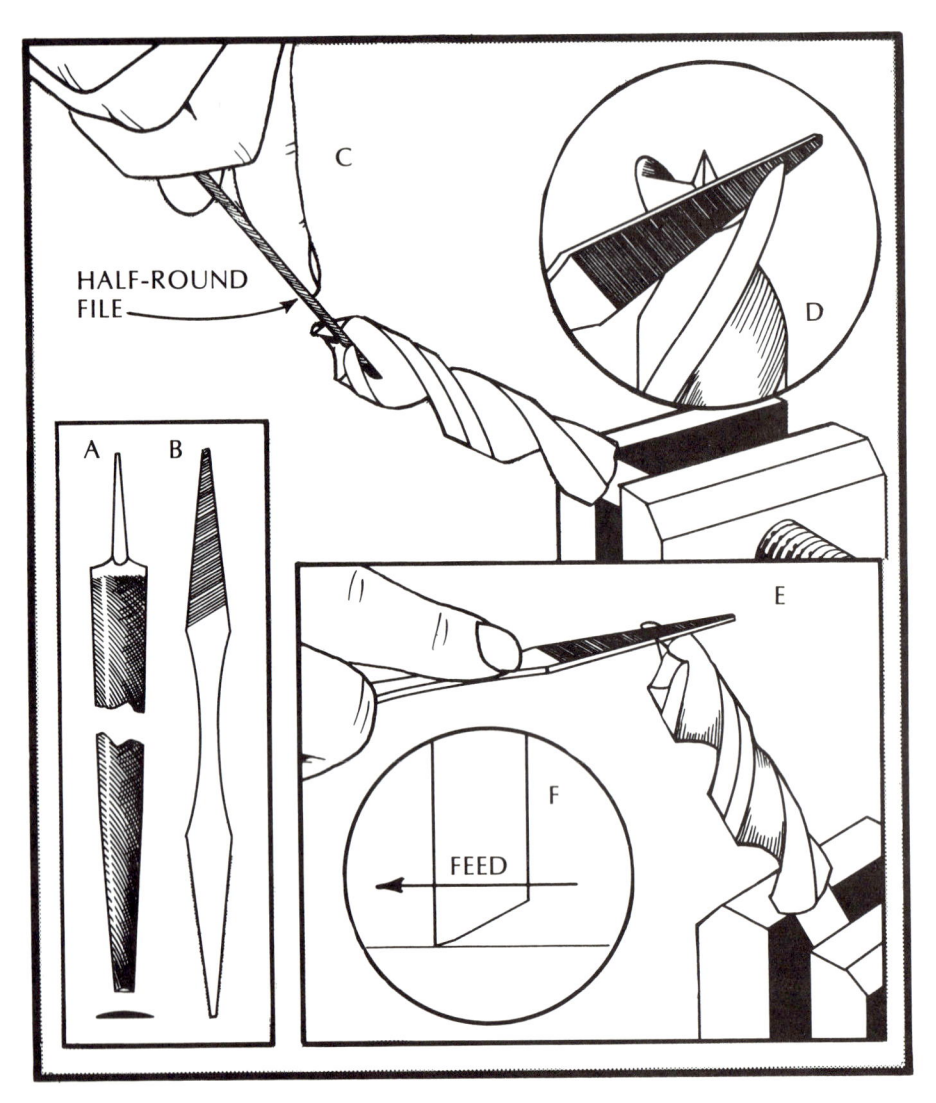

Fig. 8-16: Steps in sharpening an auger bit.

118

INCH/MILLIMETER CONVERSIONS

INCHES TO MILLIMETERS
Multiply inches by 25.4

MILLIMETERS TO INCHES
Multiply millimeters by 0.03937

INCHES	MILLIMETERS	INCHES	MILLIMETERS	MILLIMETERS	INCHES
.001	.025	17/32	13.4938	.001	.00004
.01	.254	35/64	13.8906	.01	.00039
1/64	.3969	9/16	14.2875	.02	.00079
.02	.508	37/64	14.6844	.03	.00118
.03	.762	19/32	15.0812	.04	.00157
1/32	.7938	.6	15.24	.05	.00196
.04	1.016	39/64	15.4781	.06	.00236
3/64	1.191	5/8	15.875	.07	.00276
.05	1.27	41/64	16.2719	.08	.00315
.06	1.524	21/32	16.6688	.09	.00354
1/16	1.5875	43/64	17.0656	.1	.00394
.07	1.778	11/16	17.4625	.2	.00787
5/64	1.9844	.7	17.78	.3	.01181
.08	2.032	45/64	17.8594	.4	.01575
.09	2.286	23/32	18.2562	.5	.01969
3/32	2.3812	47/64	18.6531	.6	.02362
.1	2.54	3/4	19.050	.7	.02756
7/64	2.7781	49/64	19.4469	.8	.0315
1/8	3.175	25/32	19.8438	.9	.03543
9/64	3.5719	51/64	20.2406	1.0	.03937
5/32	3.9688	.8	20.32	2.0	.07874
11/64	4.3656	13/16	20.6375	3.0	.11811
3/16	4.7625	53/64	21.0344	4.0	.15748
.2	5.08	27/32	21.4312	5.0	.19685
13/64	5.1594	55/64	21.8281	6.0	.23622
7/32	5.5562	7/8	22.225	7.0	.27559
15/64	5.9531	57/64	22.6219	8.0	.31496
1/4	6.35	.9	22.86	9.0	.35433
17/64	6.7469	29/32	23.0188	1 CM	.3937
9/32	7.1438	59/64	23.4156	2 CM	.7874
19/64	7.5406	15/16	23.8125	3 CM	1.1811
.3	7.62	61/64	24.2094	4 CM	1.5748
5/16	7.9375	31/32	24.6062	5 CM	1.9685
21/64	8.3344	63/64	25.0031	6 CM	2.3622
11/32	8.7312	1.0	25.4	7 CM	2.7559
23/64	9.1281	2.0	50.8	8 CM	3.1496
3/8	9.525	3.0	76.2	9 CM	3.5433
25/64	9.9219	4.0	101.6	1 DM	3.937
.4	10.16	5.0	127.0	2 DM	7.874
13/32	10.3188	6.0	152.4	3 DM	11.811
27/64	10.7156	7.0	177.8	4 DM	1 Ft., 3.748
7/16	11.1125	8.0	203.2		
29/64	11.5094	9.0	228.6	ABBREVIATIONS	
15/32	11.9062	10.0	254.0	MM-Millimeter(1/1000)	
31/64	12.3031	11.0	279.4	CM-Centimeter(1/100)	
1/2	12.7	1 Ft.	304.8	DM-Decimeter(1/10)	
33/64	13.0969				

Index

Accessories, 3, 4, 15-24, 27
Adapters, electrical, 25-26
   spindle, 3, 27, 29
Adjustments, 3, 4, 15-24, 27
   spindle, 26-28
Aluminum oxide sleeves, 16
Angular boring, 57-62
   drilling, 3, 85
   sanding, 100
Asbestos board drilling, 109
Auger bits, 8, 43, 117-118
Automatic center punches, 37
Auxiliary fence, 20, 40, 41, 47, 50, 55, 63, 69, 70, 103
   table, 1-2, 22-23, 47, 97, 98, 100
   tilt table, 20-21, 58, 61, 64, 102
Awl, 32, 38, 55

Bare faced tenons, 74, 75
Bar parallels, 82
Bases, 1, 3
Belt care, 31
   tension, 28
Bench drill, 1, 25
Bits, 4-9, 27, 108, 117-118
   cleaning, 9
   countersink, 9, 56, 57
   glass, 108
   sharpening, 9, 117-118
   wood, 4-9
Blind holes, 8
   mortises, 73
   wedged tenons, 74, 75
Boring deep holes, 47-49
   equal compound angles, 57, 58-60
   holes in series, 3, 50-52
   horizontal, 54-55
   in wood, 45-70
   mortises, 79-80
   round work, 52-54
   screw holes, 55-57
   simple angles, 57-58
   unequal compound angles, 57, 60-61
Brad points, 4, 5
Buffing metal, 104-105
   wheels, 17, 105
   wood, 105

Calipers, 35-36
Capacity of drill presses, 2, 3
Carbide-tipped drills, 14
Center punches, 37-38
Centering table, 29, 43, 45
Ceramic drilling, 108, 109
Chalk, 33
Chuck keys, 2, 3, 29, 43, 81
Chucks, *see geared chucks*

Circle cutters, 7, 8, 50, 69
Circular sanding, 100-102
Clamping jig, 85-86
Clamps, 18, 39, 53-54, 83, 86, 91
Clay pottery drilling, 109
Column fence, 70
   storage rack, 24
Columns, 1, 2, 3
Combination square, 33-34, 37, 72
Commercial grinding attachments, 114-116
Compound angles, 57, 58-61
Concealed haunched tenons, 74, 75-77
Cone pulleys, 2, 3, 19-20, 28, 40, 43
Contour sanding, 98-99
Coolants, 14, 88
Copper sulphate, 33
Counterboring, 9, 56, 57, 67, 92, 93-94
Countersink bits, 9, 56, 57
   plates, 93, 94
Countersinking, 9, 56-57, 92-93
Cutting mortises, 73-80

Damaskeening metal, 106
Decimal equivalents, 86
Decorative boring, 69
Deep hole boring, 49
Depth adjustment gauge, 3-4, 33, 46
   stop, 65-66
Diamond mortises, 79
Dimpling metal, 94
Disk sanding, 14-15, 103
Dividers, 34-35, 37-38
Double-spur bits, 4-5, 6
Dowel centers, 38
   holes, 38-39, 62-64
      locating of, 38-39, 62-63
      mechanical spacing of, 63-64
   joints, 65, 66
   making, 9, 66-67
   pointing, 110
   pops, 38, 39, 62, 63
Dowels, 62-66
Drawing a hole, 88
Drift keys, 29
Drill gauges, 10-11, 14, 111-112
   grinding, 112-117
      wheels, 114
   point variations, 114
   sharpening, 111-117
   storage racks, 15, 23-24, 25
   web-thinning procedure, 113-114
Drilling asbestos board, 109
   ceramics, 108, 109
   clay pottery, 109
   hard rubber, 107
   glass, 107-108
   metal, 1, 4, 9, 81-96

paper, 109
plaster, 109
problems, 90
procedures, 87-89
round work, 82-83, 84, 87-88
sheet metal, 88-91, 94
small pieces, 17-18
tapped holes, 95-96
Drills, 4, 8, 9-15, 111-117
flat drills, 110
grinding, 112-117
inserting, 28-29
twist, 8, 9-15, 111-117
Drum sanders, 14, 15, 97-98
sanding, 14-15, 97-103

Electrical adapters, 25-26
source, 25-26, 42
End boring, 40, 69-70
Equal compound angles, 57, 58-60
Expansive bit, 8, 9, 49

Feed speed, 41
Fences, 40, 41, 71, 72
Files, 17, 104, 105
Fixtures, 39-40, 82-83
Flat drills, 110
Flexible shafts, 17, 18, 104, 105
Floor models, 1, 25
Fly-cutter circle cutters, 7, 8, 50, 69
Foerstner bits, 8, 9
Fox tenons, 75

Garnet coated sleeves, 16
Gauges, 33, 36-37, 111-112
Geared chucks, 2, 3, 15, 16, 27, 28-29, 71, 81
Glass drilling, 107-108
bits, 108
Grinding drills, 112-113
guides, 113, 114
wheels, 115
Grounding the machine, 25-26

Half-inch hole spindle, 15, 16, 26, 71
Haunched tenons, 74, 75
Heads, 1, 2, 3, 28
Height gauges, 33
Hermaphrodite calipers, 35-36
High-speed drills, 14
Hold-downs, 15, 17-18, 40, 41, 71, 72
Hole saw, 6, 7, 50, 91
Holes in series, 3, 50-52
Hollow-spiral bits, 4, 6
Horizontal boring, 54-55
sanding, 102

Inch/millimeter conversions, 119
Indexing table, 18-19
Iron oxide, 33
Inserting drills, 28-29
mortising chisels, 71-72
Integral tenons, 68, 69
Interchangeable spindles, 2, 15, 26

Jigs, 39-40, 52, 54, 65, 69-70

Lamp attachments, 19-25
Lapping, 94-95
Laying out work, 31-39
Locating dowel holes, 38-39, 62-63
Lubrication, 31

Machine spur bits, 5, 8, 9, 66
Magazine rack, 60
Making dowels, 9, 66-67
flat drills, 110
plugs, 9, 67-68
Marking gauges, 33
substances, 33
Metal buffing, 104-105
drilling, 1, 4, 9, 81-96
polishing, 105-106
Metric drills, 12-13, 15
scales, 32
Millimeter/inch conversions, 119
Miter squares, 33
Mitered tenons, 74, 75
Mixing paints, 109
Models, 1, 19-20
Mortise-and tenon with splines, 74
Mortises, cutting of, 73-80
Mortising, 1, 71-80
attachments, 15-16, 71-72
bits and chisels, 15, 71-72, 79
odd-shaped work, 78-79
round stock, 77-78
spindle speeds, 72
Motors, 2, 3, 4, 27, 28, 31, 42, 72
Multiple drilling, 3, 50-52
Multi-speed attachments, 19-20
Multispur bits, 8, 9, 49

Paint stirrers, 109
Parallels, 82
Pattern sanding, 99
Pencils, 32, 33, 37
Pilot holes, 8-9
Pinned tenons, 76
Plug adapters, 25-26
cutters, 9, 10, 66, 117
sharpening, 117
fitting, 67-68
making, 9, 67-68
Pocket holes, 62
Pointer method, 87
Pointing dowels, 110
Polishing bonnets, 17, 105
Power auger bits, 8
source, 25
Profile bits, 8-9, 56, 117
sharpening, 117
Protractor heads, 34, 35
Protractors, 34, 35
Pulleys, 2, 4, 19-20, 28, 40, 43
Punches, 33, 37-38, 87

121

Quill adjustments, 28
Quills, 2, 3, 4, 27, 28, 31, 42, 72

Radial drill press, 2-3, 27-28, 43, 54-55, 58, 60, 64, 92, 102, 103
   adjustments, 27-28, 30-31
Rail joints, 76
Rasps, 17, 79, 80, 104
Reaming, 92, 94-95
Rings, 82
Riser blocks, 57-58, 59, 60, 103
Rotary files, 17, 104, 105
   hole saws, 6, 7, 50, 91
   indexing tables, 18-19
   rasps, 17, 79, 80, 104
Round tenons, 68
Rulers, 31-32, 35, 36

Safety goggles, 24, 38, 43, 105
   tips, 43-44
Sanding, 1, 16-17, 97-103
   contours, 98-99
   disk, 14-15, 103
   drum, 14-15, 97-103
      sizes, 14, 15, 97-98
   metals, 103
   miters, 103
   operations, 98-103
   plastics, 103
   rabbets, 102-103
Scratch awl, 32
Screw holes, 8-9, 55-57
Scribers, 32, 33, 36, 37
Self-ejecting chuck keys, 3, 43
   feed bits, 8, 9
Series hole boring, 3, 50-52
Shop vacuum, 24
Sharpening plug cutters, 117
   profile bits, 117
   spade bits, 117
   speed bits, 117
   twist drills, 111-117
   wood auger bits, 117-118
Side mortises, 77
Silver nitrate, 33
Single-twist solid-center bits, 4, 5, 6
Sizes, 1-2
Spacer blocks, 51, 64
Spacing punch, 37
Spade bits, 5, 6, 117
Speed-type bits, 5, 6, 117
Speeds, 2, 19-20, 28, 40 43
Spindle adapters, 3, 27, 29
   adjustments, 27

end play, 28
replacement, 26-27, 28
return spring, 29-30
speeds, 2, 19-20, 28, 40
taper, 27
Spindles, 2, 15, 25, 26-28, 42, 43
Spinning rivets, 109-110
Spot facing, 92, 94
Spring problems, 83
Straight sanding, 100
Stop bolts, 81
   pins, 50-51, 63
Storage drill racks, 23-24
Strap clamps, 82
Stub tenons, 74, 75
Surface gauges, 36-37

Table models, 1, 25
Tables, 1-2, 3, 20-23, 29, 31, 61
Tangs, 8, 14, 29
Tap styles, 96
Taper mounted chucks, 3
   shank drills, 14, 29
Tapped holes, 95-96
Tapping, 92, 95-96
Tenon joints, 73-77
   with long and short shoulders, 74, 75
Three-way joint, 76
Through-wedged tenons, 74, 75
Tilting table, 20, 47, 57, 64, 91, 103
Troubleshooting guide, 42-43
Try squares, 33
Tusks joints, 75, 76
Twin mortise tenons, 76
Twist drill parts, 10-11, 14
   points, 10
   shanks, 10, 14, 29
   sharpening, 111-117
   sizes, 12-13, 14-15
Twist drills, 8, 9-15, 111-117
   high-speed, 14

Unequal compound angles, 57, 60-61
U. S. Customary System of Units, 32, 119

V-blocks, 39-40, 47, 52-53, 77, 82-83, 84
Vises, 17-19, 39, 53-54, 84-85

Web thinning, 113-114
Whiting, 33
Wood boring, 4, 9, 45-70
   polishing, 105-106
   plugs, 9, 67-68